Astronomy from Space

Sputnik to Space Telescope

edited by
James Cornell
and
Paul Gorenstein

The MIT Press
Cambridge, Massachusetts
London, England

This book was set in Baskerville by The MIT Press Computergraphics Department and printed and bound by Halliday Lithograph in the United States of America.

Library of Congress Cataloging in Publication Data

Main entry under title:

Astronomy from space.

Includes bibliographies and index.
1. Space astronomy. I. Cornell, James. II. Gorenstein, Paul.
QB136.A79 1983 522 83-9349
ISBN 0-262-03097-7

Contents

Acknowledgments

Early in 1982, anticipating the twenty-fifth anniversary of the launch of Sputnik I, the Harvard-Smithsonian Center for Astrophysics and the Hayden Planetarium of the Boston Museum of Science, with the support of the Lowell Institute, sponsored a series of eight public lectures on the general theme of scientific results from a quarter-century of space research. Those lectures, modified, expanded, and updated by the speakers themselves and complemented by Leo Goldberg's prologue, make up this book.

"Astronomy from Space" was the sixth such lecture series presented by the Center for Astrophysics and the Museum of Science since 1976, and the second to find its way into print. (The 1980 series resulted in *Revealing the Universe: Prediction and Proof in Astronomy*, published by The MIT Press in 1982.) These lectures, in turn, grew out of a recommendation by the Planetarium Advisory Committee, a group composed primarily of Center staff members who provide scientific advice and counsel in the development of planetarium programs and displays.

The theme of space research was originally suggested by committee member Paul Gorenstein, who became one of the speakers and an editor of this volume. Planetarium Director John Carr took full responsibility for the considerable logistics involved in mounting the talks, and various members of his staff, particularly Valerie Wilcox, Matt Stein, and Ray Crane, were most helpful throughout the series.

At the Center, during the editing of talks into chapters, invaluable assistance was given in the preparation of illustrations by Joseph Singarella, John Hamwey, and Beryl Langer, in photographic work by Charles Hanson, and in manuscript preparation by Mary Juliano and Gerda Schrauwen. Anne Omundsen offered sensible and sensitive editorial comments on the original manuscripts, and John Harris provided counsel on contractual matters. Center scientific staff members Paul Blanchard, Fred Franklin, F.R. Harnden, Jr., Robert Kurucz, George

Withbroe, and Fred Whipple generously gave advice, counsel, and comment on individual chapters.

John Lowell, Director of the Lowell Institute, provided the generous support that has made the now-annual Lowell Lectures on Astronomy an important part of the Boston Museum of Science's public education program.

Astronomy from Space

1

Prologue: Astronomy Before the Space Age

Leo Goldberg

When I began working in astronomy, in 1933, astronomy was almost strictly an observational science with very little understanding of the physical meaning of observations. A few brave souls were just beginning to apply the new discoveries of atomic physics to astronomy, but their work was viewed with a certain amount of skepticism. Nuclear physics as a science hardly existed. Henry Norris Russell had recently made the first determination of the chemical composition of the sun, and Sir Arthur Eddington and others were insisting, although without proof, that the energy needed to keep the stars shining for millions and billions of years must come from the transmutation of hydrogen into helium. Fifteen years earlier, Lord Rutherford had transmuted nuclei by bombardment in the laboratory. Eddington said "What is possible in the Cavendish Laboratory may not be too difficult in the sun," and suggested that his critics, who thought the interiors of stars were not hot enough for the conversion of hydrogen into helium, "go and find a hotter place."

The real nature of the sun was a total mystery. The sunspot cycle had been known for over a century, and the magnetism of sunspots had been detected over 20 years earlier, but no one had any idea how magnetic fields were generated or what powered the sunspot cycle. Astronomers were struggling to learn the identity of the mysterious set of spectrum lines radiated by the solar corona, which could then be observed only during the few moments of a total eclipse. The overwhelming majority believed that the temperature of the corona was about 5,000°K. They would have been astounded to be told (as they were in 1942) that the lines originated from very highly stripped atoms signaling a temperature of more than a million degrees.

The moon and the planets were hardly being studied at all. It was not that astronomers did not recognize the importance of these objects as possible abodes for life and as repositories of clues to the origin of the Earth and the solar system; the blurring effects of the Earth's atmosphere so handicapped the vision of even the largest telescopes that there was simply no hope of observing planets with sufficient clarity to answer important questions.

We knew from the work of Harlow Shapley and others that our sun was a somewhat undistinguished star—one of some 300 billion grouped in the gigantic, flattened disk of a galaxy 100,000 light years in diameter—located about two-thirds of the way to one edge of the galaxy. We suspected that our galaxy, the Milky Way, had a spiral shape similar to that of many other galaxies that filled the universe and could be photographed with telescopes looking out through the spaces between the stars in the Milky Way. However, the shape of our own galaxy seemed forever hidden from view by great clouds of dust in interstellar space. No one suspected at that time that the recently discovered radio waves from the galaxy would later become the means of exploring the galaxy in detail because the dust is transparent to radio waves.

We also knew that the observable universe was populated by billions of such galaxies, separated from one another by distances on the order of a million light years and extending out in all directions as far as the telescope could see. We had recently learned from Edwin Hubble's work with the great 100-inch telescope that the universe is expanding, with galaxies rushing away in all directions; the farther away the galaxy, the faster it is receding. The speed of recession was thought to be increasing at the rate of 100 kilometers per second per million light years. It was as though the universe had started with a big bang, with the fastest-moving parts having traversed the greatest distances. Knowing the speeds and the distances, we readily calculated that the expansion had begun about 3 billion years ago. This has now been revised upward to 10–20 billion years.

As seen 50 years ago, the universe was a quiet and peaceful place, slowly evolving and exhibiting little of the violent and explosive behavior that we now take for granted. The intervening years have seen astronomy expand like the universe itself, driven by a sequence of technological and scientific advances and by a tenfold growth in the number of astronomers. Looking backward, one now realizes that the period of rapid growth actually began around the turn of the century, but that the pace of discovery speeded up dramatically after World War II under the impetus of wartime technology and then fairly exploded with the coming of space vehicles for exploration and observation.

Just before the last century ended, the great telescope builder George Ellery Hale founded the Yerkes Observatory of the University of Chicago and built its majestic 40-inch refractor. Moving on to California, he established the Mount Wilson Observatory and created the 60-inch, 100-inch, and 200-inch reflecting telescopes and the 60-foot and 150-foot towers for observation of the sun. Long before the completion of the 200-inch telescope, work with smaller reflecting telescopes had laid the groundwork for a conceptual model of the universe. That model proceeded from the "big bang" hypothesis, which was already in place by 1930 and which still serves as a guide for most modern astronomical research.

In the 1920s there had been a revolution in our understanding of how atoms and molecules emit radiation and interact with one another under any given set of physical conditions. Astronomers could then begin to analyze astronomical spectra and to infer chemical compositions, temperatures, densities, motions, and magnetic and electric fields in the atmospheres of stars, in the interstellar medium, and in nearby galaxies.

One of the first fruits of the new astrophysics came in 1928 with the first analysis of the chemical composition of a star, in this case the sun, by Henry Norris Russell at Princeton. His former student Donald H. Menzel, who came to Harvard in 1932, made me keenly aware of the importance of determining the chemical compositions of stars as a means of getting an understanding of stellar evolution. While still an undergraduate, I was put to work calculating certain parameters of atomic physics—transition probabilities, they are called—which are fundamental to the derivation of composition. Menzel's vision was not fully shared by his colleagues at Harvard; the director of the observatory, Harlow Shapley, once suggested to me that, in view of my research interests, I ought to consider transferring to the physics department or to the Massachusetts Institute of Technology. Since that time, the stars, once thought to be uniform in makeup, have been found to differ widely in chemical composition, and this knowledge has been absolutely decisive in testing theories of stellar evolution.

Another vital development in the understanding of stellar evolution had been the application of nuclear physics. In the late 1930s, the physicists Hans Bethe, C. L. Critchfield, George Gamow, and C. F. von Weizsäcker discovered the detailed nuclear processes by which stellar energy is generated in the interiors of stars through the synthesis of hydrogen into helium. It also seemed plausible that most of the helium in the universe was produced in the first three minutes after the "big bang" by the successive capture of neutrons; however, for a number of years the processes by which heavier elements were built up from helium remained a mystery. In the late 1940s, Gamow and his students R. H. Alpher and R. C. Herman proposed that the heavy elements were also produced in the hot, early universe by the sucessive capture of neutrons. This theory soon foundered upon the realization that when helium, with a mass of 4 (two protons and two neutrons), captured a neutron, the resulting nucleus of mass 5 would be unstable and would immediately disintegrate. Without mass 5, there could be nothing heavier.

This problem was resolved by Ernst Öpik and Edwin Salpeter, who showed independently that the mass-5 difficulty could be circumvented by the fusion of three helium nuclei to form carbon, but that these and other heavy-element-creating reactions could take place only in stellar interiors and not in the early universe (as Gamow proposed). Fred Hoyle was among those who argued that heavy elements could be formed in stellar interiors, which led to some lively exchanges between him and Gamow.

In 1956, Hoyle, William A. Fowler, Margaret Burbidge, and Geoffrey Burbidge worked out in detail the processes by which heavy elements are "cooked" in stellar interiors and set the stage for the picture of stellar evolution that now serves as a working model for all studies of the birth, life, and death of stars. New stars of all masses up to about 100 times the mass of the sun are believed to condense out of dense clouds of dust and gas in the interstellar medium. During most of their lives, stars radiate energy liberated by the burning of hydrogen. In their infancy they are powered by gravitational contraction, but in old age they get their energy by fusing helium into carbon, oxygen, and so on into still heavier elements. The more massive the star, the faster it radiates and the shorter is its life, from a few million years for the most luminous ones to more than 10 billion years for stars less massive than the sun. After the hydrogen is exhausted, the stars swell up enormously and their outer layers cool off as they become red giants and supergiants, variable stars, and generally unstable objects. After all the nuclear fuel is gone, stars weighing only a little more or less than the sun contract slowly, cool off, and die peacefully as white dwarfs. More massive stars collapse on themselves and explode as supernovas, hurling most of their matter into space and leaving behind a superdense neutron star or even a black hole. One supernova will explode every hundred years on the average in each galaxy.

Our galaxy is probably at least 10 billion years old, and during that time many generations of stars have been born and have died. Since heavy elements are made in stars and are cast out by mass loss and supernova explosions, the gas and dust from which new stars form has gradually acquired a higher and higher percentage of heavy elements relative to hydrogen. In fact, very young massive stars seem to have higher concentrations of heavy elements than old stars like the sun, and certain classes of stars are found to have more carbon or more nitrogen or more strontium—all for reasons having to do with nuclear transmutations in stellar interiors during evolution. And so, while the same chemical elements are found everywhere in the universe, their proportions may differ considerably, and chemical composition is the most important clue to the processes of stellar evolution.

Until about 1946, everything that was known about the universe had been learned from radiation contained within a small strip of the electromagnetic spectrum: the narrow band visible to the human eye, its near-ultraviolet and near-infrared extensions, and a few relatively narrow segments of the far-infrared spectrum. All wavelengths shorter than about 3,000 Ångstroms (1 Å = 1/100,000,000 centimeter)—which means most ultraviolet radiation and all x rays and gamma rays—are blocked by the Earth's atmosphere, as are most infrared rays. Radio waves of frequency lower than about 20,000 hertz (a hertz is one cycle per second) are also reflected back into space by the Earth's ionosphere. It was known that radio waves of higher frequency, with wavelengths less than about 20 meters, pass easily through the atmosphere. But such radiation was not thought to be of astronomical importance, even after 1932, when Carl Jansky, an engineer at Bell Laboratories, published a paper reporting the accidental detection of radio waves from the Milky Way. Astronomers are naturally cautious and conservative; besides, had not theoretical calculations proved that thermal radio emission from the galaxy would be too feeble to be detected? Neither the widespread occurrence of violent events in the universe nor the strongly nonthermal character of their radiation was yet appreciated by most astronomers. It took the development of radar—and groups of enterprising radio physicists and engineers in Australia and Great Britain who knew too little about the conventional wisdom of astronomy to be turned away—to uncover the riches of the radio spectrum.

At about the same time, the first steps were being taken to break through the atmospheric wall blocking the observation of the remainder of the spectrum. In the 1920s and the 1930s many unsuccessful attempts

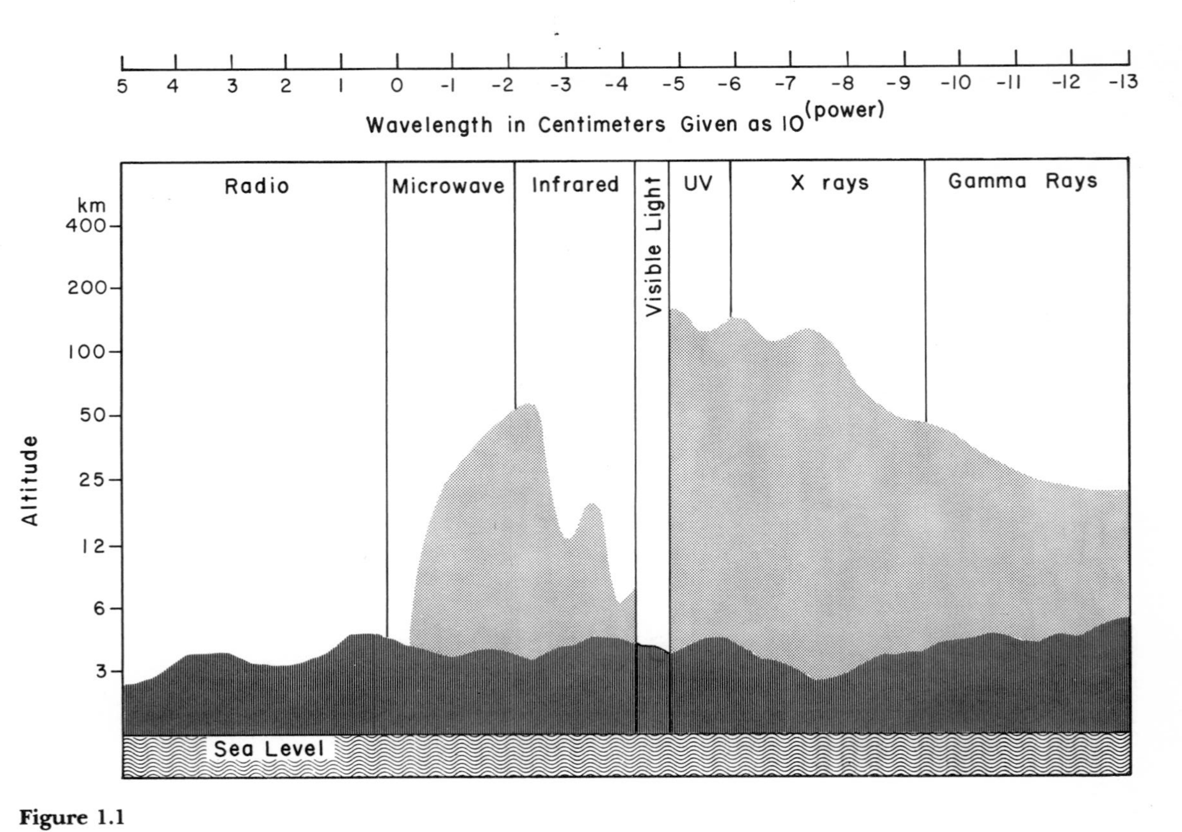

Figure 1.1
The transparency of the Earth's atmosphere to electromagnetic radiation from space. (Smithsonian Astrophysical Observatory illustration by Joseph Singarella)

were made to detect ultraviolet radiation from ground stations in the far north (where the ozone layer was conjectured to be relatively thin), from the tops of mountains, and from manned balloons. It was not until late in World War II, when the Germans developed the V-2 rocket, that the means for climbing above the atmosphere were acquired.

Beginning in late 1946 with captured V-2s and continuing with newly designed sounding rockets in the United States, Canada, Europe, and Australia, several groups of scientists mapped the solar spectrum from short x-ray to near-ultraviolet wavelengths. Among other things, they confirmed the high temperature of the corona, demonstrated the huge scale of energy release during solar flares, and paved the way for the satellite observatories that were soon to come.

The orbiting of a satellite by the Soviets on October 4, 1957, had a stunning effect on astronomy. Both the United States and the U.S.S.R. had announced intentions of launching satellites during the International Geophysical Year of 1957–58, but the sheer size of the 84 kilogram Sputnik I was a great surprise. Of course, we soon realized that large military rockets were capable of orbiting whole observatories and sending instrumented probes to the moon and the planets. In June 1958, the U.S. Advanced Research Projects Agency estimated that by 1960 payloads up to 3,000 pounds could be put into Earth orbit and that others of 600–1,000 pounds could be sent to the moon or the planets. Scientific satellites weighing 25,000–50,000 pounds were anticipated for the mid-1960s, and it was predicted that the multiplexing of rockets could soon increase the size of possible orbital payloads to 100,000 pounds.

Having been led to look forward to payloads of no more than 120 pounds, the scientific community simply was not prepared to think in terms of such massive equipment. To encourage the widespread participation of scientists in the space program, the National Academy of Sciences, in June 1958, constituted a Space Science Board under the chairmanship of Lloyd V. Berkner, who invited me and fourteen other scientists to join. One of the board's first actions was to send some 150 telegrams to members of the scientific community soliciting proposals for satellite experiments. Over a hundred replies were received, and the proposals were evaluated by a number of committees; as a result, when the National Aeronautics and Space Administration came into existence in October 1958 it already had available the nucleus of a solid space-science program. It is interesting to read these proposals now, after 25 years, and to compare the expectations with what actually happened.

Figure 1.2
Earth's natural satellite, the moon, shares the predawn sky with the third-stage rocket of Sputnik I in this photograph taken by an amateur astronomer in Arlington, Massachusetts, on October 16, 1957. (Smithsonian Astrophysical Observatory photo)

Observatories in space were expected to offer the following advantages:

- Detectors could be placed far above Earth's atmosphere, which totally absorbs and blocks from view all wavelengths of electromagnetic radiation shorter than about 3,000 Ångstroms and a large fraction of infrared radiation. Radio waves of frequency lower than about 20,000–30,000 hertz are also reflected back into space by the ionosphere.
- A perfect optical telescope, operating in the near vacuum of space, will form an image of a point source whose diameter is inversely proportional to the aperture of the telescope. For a 200-inch telescope, the image in visible light can be as small as 0.02 second of arc, which is about the angle subtended by a dime at a distance of over 100 miles. By contrast, on Earth, where light must pass through the turbulent atmosphere, a celestial source will be puffed up to a diameter of at least 0.5 second of arc under the best conditions and

to several seconds of arc when the "seeing" is at its worst. The spreading of the image destroys definition and greatly lengthens the time required to record star images. (The importance of imaging the sun with high resolution had just been demonstrated by photographs taken with a 36-inch telescope from a high-altitude balloon in Princeton's Project Stratoscope.)

- Ground-based telescopes must detect faint objects against the glare of light emitted by atoms and molecules in the upper atmosphere. A telescope in orbit above the airglow sees a relatively dark background arising only from the feeble light of unresolved stars in the Milky Way and from sunlight scattered by dust particles in the ecliptic plane of the solar system (the so-called zodiacal light). Elimination of atmospheric turbulence and airglow can increase the distance of faint galaxies reached by a large telescope by a factor of 10 or more.
- It is much easier to support and maintain the precise parabolic figure of a large telescope mirror in the gravity-free environment of space than on Earth. This problem can become critical for mirrors over 200 inches in diameter.
- Instrumented probes, with or without human occupants, could be sent to explore other bodies of the solar system, make *in situ* measurements, and return close-up pictures and samples to Earth.
- Finally, the weather in space is always perfectly clear, and when a telescope is in a stationary, 24-hour orbit its observations are not interrupted by sunrise or sunset.

Since the launching of Sputnik I the vast majority of the projects that could be envisioned in 1958–1960 have been accomplished or are well on the way to fruition. There have also been many surprise discoveries that have radically altered our view of nature and forced major changes in the goals and priorities of astronomy. The initial expectations for astronomy from space, and how they were met, are the subjects of the remainder of this chapter.

Earth

Geodesy, meteorology, and the investigation of charged particles and magnetic fields in the Earth's upper atmosphere and into the interplanetary medium were high on the agenda of space science in 1958. The motion of an Earth satellite is controlled by the gravitational attraction of the Earth. The orbit of the satellite can be computed on the assumption that the Earth is a perfect sphere. Accurate tracking of a geodetic satellite was expected to reveal departures from sphericity and to yield distances between widely separated points on Earth to an accuracy of 30 meters. Earth was found to be pear-shaped, and by 1986 the error of distance determinations is expected to be as little as 2 centimeters.

Meteorologists had long been aware of the power of synoptic pictures of weather events as an aid to forecasting, and knew the importance of making observations as high as possible in the atmosphere. By 1958 they had progressed from manned balloons to kites to airplanes to balloon-borne radio sounding to large sounding rockets. Still, despite communication with a worldwide network of observation posts, less than 20 percent of the mass of the atmosphere could be probed, and large storms over uninhabited regions might remain undetected for days. The Space Science Board strongly recommended an appropriate global distribution of meteorological satellites to keep a continual watch on "every storm on Earth," to "note the birth of new storms and the death of old ones." This goal has long since been achieved, and weather forecasting has been accordingly transformed from a guessing game to a science.

Interplanetary space

Before the development of satellites and space probes, our only knowledge of conditions in the Earth's outermost atmosphere and in interplanetary space was that deduced from studies of the airglow, auroras, the zodiacal light, the outer solar corona, comet tails, and meteors. All we knew about charged particles and magnetic fields had been inferred from the behavior of cosmic rays and from the effects of the sun on the Earth. The study of particles and magnetic fields in space was galvanized by the very first successful U.S. satellite, Explorer I, and its discovery of the Van Allen radiation belts in January 1958. By October of that year, the Space Science Board had prepared a comprehensive plan for the initial exploration of space around Earth and far into the interplanetary medium.

The SSB program was concerned with magnetic fields, low-energy particles, cosmic rays, x rays, auroras and airglow, and the detection of interplanetary particles. Measurements of particles and fields were to be carried out in the vicinity of Earth, in the interplanetary medium, near the moon, and around other planets (especially Venus and Mars). (The exploration of more distant planets, such as Jupiter and Saturn, was not seriously considered until 10 years later.) The cosmic-ray program envisioned the measurement of composition and intensity, and their variations with time, of solar and galactic cosmic-ray particles, as well as searches for antimatter, neutrons, magnetic monopoles, and gamma rays. Although the SSB had received only one proposal for x-ray experiments (from scientists in the Signal Corps of the U.S. Army), it urged the development of experiments to produce x-ray images of the sun and of solar flares and to map the celestial sphere in the light of x rays. At that time, nonsolar x rays had not been detected; solar x rays had been studied intensively by rocket, but without resolution of the solar image.

The particles-and-fields program covered a lot of territory, and dozens of satellite missions were needed to carry it out. A variety of spacecraft have been used, notably the OGOs (Orbiting Geophysical Observatories), POGOs (Polar Orbiting Geophysical Observatories), Explorers, IMPs (Interplanetary Monitoring Platforms), Pioneers, Mariners, and, most recently, the two Voyagers.

Two of the earliest discoveries that stand out in my memory are the direct observation of a high-speed solar wind and the determination of the sector structure of the interplanetary magnetic field. The existence of a wind blowing outward from the sun at a speed of several hundred kilometers per second had been deduced from the orientation of comet tails by Ludwig Biermann and shown theoretically by Eugene N. Parker to be a consequence of the hot solar corona. Magnetic fields may have either positive or negative polarity. Monitoring of the interplanetary field associated with particles flowing out of the sun and arriving at the Earth revealed that at any one time the field associated with the plasma is of one polarity, either plus or minus, but that the polarity changes abruptly every few days. Particles flow out of the sun in a radial direction, and only those from the center of the solar disk as seen from the Earth will reach the Earth. But the sun rotates once in about 25 days, and therefore particles received at the Earth originate from a constantly changing area of the sun. Evidently, the sun's equator is divided into about six sectors such that the polarity of the magnetic field associated with ejected gas alternates in successive sectors. The reason for this behavior is still not known.

The moon and the planets

From the outset, the exploration of the solar system drew the interest and participation of researchers from physics, geophysics, chemistry, applied mathematics, biology, and engineering, as well as from astronomy. No other human activity can boast such a marshaling of all of the resources of the natural sciences. One of the great triumphs of this extraordinary effort has been the close-up examination of the moon and the five nearest planets and their satellites. Men have landed on the moon, instruments have been deposited on the moon, Venus, and Mars, and samples have been brought back from the moon. The moon is close enough to the Earth that its surface can be looked at in considerable detail through a telescope of moderate size under good observing conditions, but there was no way of guessing the structure, composition, and age of the surface material. The composition of planetary atmospheres could be deduced in part from spectroscopic observations, especially by infrared radiation, but little else could be learned from the dancing, blurry images that astronomers saw through their telescopes. In June 1958, Harold C. Urey was appointed chairman of an SSB committee concerned with planning missions for exploring the moon and the planets. The program was ready in summary form within five months, and although limited to the moon, Venus, and Mars it proved to be a remarkable forecast of what occurred in the next 15 or 20 years. The goal of the program was no less than to gain an understanding of the nature, origin, and evolution of our planetary system and of life within it. The committee identified dozens of areas of inquiry, including the density and composition of the moon's atmosphere; the composition and structure of the surfaces and the interiors of the moon, Mars, and Venus; the temperature distribution within the moon; the exact mass of Venus; the compositions, thicknesses, densities, and temperature distributions of the atmospheres of Venus and Mars; the composition of the clouds of Venus; the rotation period of Venus; the presence or absence of oceans on Venus; the presence or absence of dormant life spores on the moon's surface; and the presence or absence of life forms on Mars and Venus.

The moon

Since the moon has little or no atmosphere, its surface has not suffered erosion and must be full of information about its history. However, the view from Earth allowed all sorts of conflicting theories as to the nature of the surface and how it was formed. Before the Apollo missions, there was vigorous debate as to whether the lunar surface was once completely melted, whether the maria were lava flows from the interior or dust bowls, and whether rocks differentiated by melting were found on the surface. The oldest of all questions was whether the craters were predominantly of meteoric or volcanic origin. Another key question was the time of formation of the moon's surface, which cannot be directly determined from Earth but which could be resolved by employing the potassium-argon method on different parts of the surface.

The kinds of missions proposed for the lunar program included a satellite of the moon (Lunar Orbiter), a hard landing (Ranger), a soft landing (Surveyor), an unmanned landing with return of samples, and a manned landing (Apollo). The possibility of a manned landing on the moon was clearly in the minds of the scientific planners from the start, but a certain lack of enthusiasm was strongly implied by the inclusion of an instrumented landing with return of samples, which was deleted after the manned landing was decided upon. (This objective was pursued by the Soviet Luna missions, however.)

There was certainly a clear consensus among scientists in 1958 that the scientific objectives of lunar exploration could be achieved without manned landings. However, man in space proved to have more than symbolic value—if not during the Apollo missions, then certainly in Skylab, which would have been a disastrous failure without men on board to repair damage suffered during the launch, open stuck camera shutters, and perform vital observing tasks. Moreover, the tools of propulsion, guidance, and communications needed for the later missions to Jupiter and beyond were forged in the heat of the national effort to land men on the moon, and might otherwise have been developed much more slowly.

Although the motivation for the Apollo mission was not primarily scientific, the scientific returns were very great. We still do not know how the moon was formed (whether by accretion of planetesimals or by fission from the Earth), but our ideas about the early history of the moon have been changed radically. Before Apollo, Earth scientists generally agreed that the moon and the planets had probably formed in a cool condition and that their interiors had gradually been warmed by the decay of radioactive elements. It is now believed that very early in its existence the moon was completely covered by an ocean of hot magma several hundred kilometers deep, and that all the inner planets (Earth included) began in the same way. Another concept that has emerged is that nearly all of the circular maria on the near side of the moon were formed in a very short time by a fierce bombardment of large planetesimals 3.9–4.0 billion years ago. A further surprise has been the realization that the same minerals, rocks, and geologic processes occcur both on the moon and on the inner planets—they are all fundamentally alike and originated in much the same way.

Venus and Mars

The best guess in 1958 was that Venus had, or once had, surface water and volcanic activity similar to that on Earth and therefore probably suffered extensive erosion. Mars was also conjectured to have had oceans, and possibly volcanic activity. Otherwise, these planets were thought to differ in detail from Earth because they were formed with different masses, at different distances from the sun, and with different amounts of volatile substances. There were speculations that Venus might be covered with oceans and that its atmosphere might contain clouds of hydrocarbons, water, or dust. The nature of the surface markings of Mars was unclear, but no one believed any longer they could be canals constructed by intelligent beings. The exploration of these planets was expected to throw light on their origin, but far and away the most exciting reason for going to the planets was to search for evidence of life. Mercury was excluded from the earliest exploration plans because its closeness to the sun was expected to cause difficulties with electrical circuitry, but these were later overcome. Mariner 10's flyby of Venus and Mercury and the many other missions to Venus and Mars by the U.S. and the U.S.S.R. have brought monumental increases in knowledge of the inner planets. Indeed, we are beginning to understand the forces that have shaped these planets, and we know approximately when these forces were at work.

Some 80 percent of the surface of the moon was formed in the moon's first 600 million years. By contrast, 60–70 percent of the Earth's surface appears to have been formed within the last 200 million years. Mars is an intermediate case, and suggests that small planets have a high ratio of surface area to volume and therefore cool off more rapidly by conduction than the larger ones. Venus is still a great puzzle, but is shaping up as the key to the understanding of planetary evolution and as an object of special interest for future planetary missions. No positive evidence of life forms on the moon or the planets has yet been found, although a very careful search was undertaken with the Viking lander on Mars.

Jupiter and Saturn

In 1958–59, the great distances to the giant planets seemed to bar their exploration in the near future. As early as 1970, however, Pioneer spacecraft were being readied to probe Jupiter's atmosphere and possibly to fly on to one of the outer planets. Jupiter was regarded as the most rewarding objective among the giant planets, by virtue of curiosity as to its magnetic field and radiation belts, its composition, the physics, chemistry, and dynamics of the lower atmosphere, and the mysterious modulation of the planet's radio emission by the motion of the satellite Io. It was also noted at the time that an unusual alignment of the planets between 1975 and 1980 would permit a spacecraft to swing by Jupiter and use its gravitational pull to go to Saturn and Uranus, which in turn would propel the craft to Neptune and Pluto. The Grand Tour concept, as it became known, was later scaled down as too costly, but it still evolved into the strikingly successful Voyager missions. The Voyager and the earlier Pioneer spacecraft have collected rich stores of information about the atmospheres, magnetospheres, and satellites of Jupiter and Saturn. The observation of the rings and satellites of both planets has been particularly exciting. The massive volcanic eruptions on Io were totally unexpected, as were revelations of tectonic activity in the crusts of other satellites. Still, although we have gained an enormous amount of information about the major planets and their satellites, we are far from understanding what we observe. The Galileo satellite, which is planned to orbit Jupiter in a few years, may solve some of the puzzles, but much more exploration is needed before we can answer the fundamental questions of planetary origin and evolution.

The sun

In the pre-Sputnik years, solar astronomers asked what caused the sunspot cycle, how flares originated, and how the solar corona could be heated and maintained at temperatures greater than a million degrees despite the fact that the temperature of the visible surface below is a relatively cool 6,000°K. We are still asking the same questions today; however, the number of possible answers has shrunk, and wholly new ways of looking at the sun have been devised. In 1958, rocket observations had confirmed that, as expected, most of the energy from the million-degree-plus corona, from solar flares, and from the active solar regions was being radiated in x-ray and ultraviolet wavelengths. Hence, the highest priority was assigned to observations of short-wave radiation, including gamma rays, x rays, and ultraviolet.

The NASA philosophy in these early planning days of space astronomy was to design a single space platform that would meet the needs of both solar and nonsolar astronomy. Indeed, at one time the agency hoped to mount both solar and stellar telescopes in the same spacecraft. After this scheme was shown to be impractical, NASA assigned the fourth spacecraft in the series of Orbiting Astronomical Observatories (OAOs) to solar research. It soon became apparent that several years of development would be required before useful solar instrumental payloads of 700 pounds could be designed. In the meantime, NASA wisely decided to proceed with a series of Orbiting Solar Observatories (OSOs) that would carry about 70 pounds of experiments in a pointed section and a number of smaller instruments in a rotating wheel, so that each experiment would look at the sun once every two seconds and scan a strip of the sky for the remainder of the time. The pointed section was stabilized to an accuracy of about one minute of arc about two axes only and was free to rotate very slowly about the line of sight to the sun.

By building on the experience with the OSOs, NASA expected that it would be possible in about five years to design an "ultimate" solar observatory satellite, equipped with a variety of telescopes and instruments capable of pointing at small features anywhere on or around the sun with an accuracy of one second of arc. The instruments would be designed to acquire x-ray, gamma-ray, and ultraviolet spectra of very high spectral resolution, as well as images of the sun in monochromatic radiation of any desired x-ray or ultraviolet wavelength and pictures of the corona in white light extending to distances of several radii from the edge of the sun. A bolometer to measure and monitor the constancy of total solar radiation to an accuracy of 0.1 percent was also envisaged.

In February of 1959 I was visited at Ann Arbor by Homer Newell, soon to become director of NASA's Office of Space Science, who asked whether I would be willing to undertake the preparation of the 700-pound experimental package for solar physics in the fourth OAO. My first reaction was shock. Rockets were wont to collapse in a column of smoke and flame on the launching pad, and no sane astronomer would think of gambling several years of his career on such a risky venture. I had never built a space instrument, and even the experts were not sure how to design equipment to operate for long periods of time in space. It took me about a minute to think of all the reasons for saying no before I said yes. If the risks were great, so were the potential rewards. In retrospect it was a foolhardy undertaking, but I have never regretted it.

As it turned out, the original plan for a "solar OAO" was abandoned, and we solar experimenters settled down to the more realistic task of instrumenting the smaller OSOs. In 1960 I moved from the University of Michigan to Harvard, where my group designed and built a series of devices of progressively greater capability that flew in OSO-4, OSO-6, and Skylab. My interest was in obtaining high-definition images of any part of the sun in the monochromatic radiation of any desired ultraviolet wavelength. By recording images in the radiation emitted by ions in various stages of ionization (for example neutral hydrogen, singly ionized helium, doubly ionized carbon, quintuply ionized oxygen, and fourteen-times-ionized iron), we were able in effect to study the properties of the solar plasma at different heights in the atmosphere as the temperature increased from 6,000°K at the visible surface to 2,000,000°K in the corona.

All of the measurements originally planned for the "ultimate solar observatory" were eventually made with the aid of a progression of spacecraft and instruments of steadily improving pointing accuracy and resolution, including eight OSOs, the Apollo Telescope Mount in Skylab, and the Solar Maximum Mission.

We are beginning to comprehend the crucial role of the magnetic field in shaping the structure of the corona and possibly in its heating. Monochromatic images of the corona projected onto the disk of the sun help us to separate structures along the line of sight, and thus to disentangle the complex geometry and dynamics of the sun's outer atmosphere. The sun looks entirely different today than it seemed to us even 10 or 15 years ago. Most of this new knowledge has been gained from space, but an entirely new property of the sun has been discovered with telescopes on the ground. These observations have revealed that the sun is a giant oscillator, vibrating like a crystal sphere in a wide range of frequencies and spatial scales. These vibrations (which serve as probes to the interior of the sun, like seismic soundings of the Earth) are telling us much about the sun's interior circulation. The value of this technique will be greatly enhanced when the measuring instruments can be placed in Earth orbit.

Beyond the solar system

Pointing a space telescope at a star is enormously more difficult than sighting at the sun, and for that reason very few rocket observations of stars, nebulas, the interstellar medium, or galaxies had been attempted up to 1958. Three-axis stabilization systems for astronomical rockets were just beginning to be developed. Some measurements had been made with unstabilized rockets recording ultraviolet radiation from whatever direction in the sky the rocket happened to be pointing in as it rolled, pitched, and yawed during its flight. Afterward, objects whose radiation had been intercepted were identified from the record of the rocket's trajectory. Under the circumstances, astronomers and NASA were bold indeed to plan immediately for a large, stabilized platform that would support telescopes up to 36 inches in aperture with a pointing accuracy of a fraction of a second of arc. This concept materialized as the Orbiting Astronomical Observatory, two of which were successfully flown, the second being the famous Copernicus Observatory launched in 1973.

In the absence of preliminary rocket experiments, the scientific goals of nonsolar ultraviolet astronomy were rather vague in the planning stages. It was known that the hotter stars, with temperatures of 10,000°K or more, emit most of their radiant energy in the form of ultraviolet radiation, and that gaseous nebulas also radiate profusely at short wavelengths. The chemical abundances of such common elements as carbon, nitrogen, and oxygen could be derived with much greater precision from the ultraviolet spectra of hot stars than from visible spectra. Hydrogen in molecular form was expected to be an abundant constituent of the interstellar gas, but it does not reveal itself in visible light. It was known to emit and absorb strong spectral lines in the far-ultraviolet range, near 1,100 Ångstroms, and its detection and measurement was a prime objective for the OAO. In fact, the molecular hydrogen radiation was first discovered through a rocket experiment, but that did not detract from the importance of the subsequent satellite investigations.

Because of the relatively primitive instrumentation and the lack of a well-defined program, astronomers recommended that galactic and extragalactic research from satellites be developed in stages, beginning with ultraviolet sky maps from small satellites with relatively crude pointing and culminating with the establishment of larger, accurately controlled telescopes in space. The NASA engineers were confident, however, that the first OAO would be ready to fly the survey programs as early as 1963 and that therefore smaller satellites would be superfluous. As it turned out, the first OAO—carrying experiments for Wisconsin and the Smithsonian Astrophysical Observatory—was not operative until 1968, but it achieved most of its scientific goals. Our knowledge of the structure and the chemical composition of the interstellar medium has increased immeasurably, and mass loss from hot stars, in the form of hot, outward-blowing winds, has been found to play a vital role in stellar evolution.

Probably the most spectacular and important product of space astronomy has been the discovery and study of x rays from outside the solar system. Until 1962, the promise of x-ray astronomy was somewhat speculative. Calculations showed that if all stars radiated x rays with the same intensity as the sun, the flux at Earth would be too weak to be detected with existing instruments. Nevertheless, some astronomers pointed out that much greater levels of x-ray emission might emanate from old, exploded stars, galaxies in collision, stellar flares, and other objects emitting nonthermal radiation. The Astronomy Committee of the Space Science Board did recommend that all-sky surveys for x rays and gamma rays be undertaken for discovery purposes, but the program was given low priority by NASA.

The first discovery of an x-ray source in the Milky Way was made by accident. Bruno Rossi and Riccardo Giacconi had been developing an x-ray telescope with the conviction that somewhere in the sky there would be sources to observe. Unsuccessful in securing support to search for x-ray sources outside the solar system, they instrumented a rocket, with U.S. Air Force sponsorship, to measure the x rays from the moon that were expected to be generated by the bombardment of the surface with charged particles from the sun. The experiment was successful, and, moreover, as the instrument scanned the sky in the process of acquiring its lunar target, its detector recorded x radiation in the constellation Scorpius from a source now known as Sco X-1.

The future

In prospect for the next decade is the Space Telescope, with which astronomers may "look back in time" and view the universe as it appeared billions of years ago. Other magnificent new space facilities are now in the planning or preparatory stage: a large optical solar telescope, an infrared telescope, a gamma-ray observatory, and an advanced x-ray telescope. On the ground, astronomers hope to build at least one enormous optical telescope in the 10–15-meter-diameter range. Radio astronomers are proposing an array of radio telescopes that will span the United States and achieve a spatial resolution equivalent to that of a single telescope 3,000 miles in diameter, or about 0.0003 second of arc. By contrast, optical astronomers are now limited to a resolution of 0.5 second on the ground, and the Space Telescope is expected to achieve 0.1 second. The resolution that can be attained by a telescope of a given size is inversely proportional to the wavelength. To gain a 0.001-second resolution in visible light, we would need an array of mirrors about 200 meters across. Because of complications introduced by Earth's atmosphere, such an array would have to go into space. This probably could not be realized before the end of the century. However, such a giant leap in optical resolving power—by a factor of 1,000—would clearly be a new beginning for astronomy, and would prove once again that the oldest of the sciences is constantly being reborn.

2

Geology of the Inner Planets

James W. Head III

Over the last two decades we have been witness to extraordinary views of planetary panoramas: Earthrise from lunar orbit, the wind-swept red deserts of Mars, the cloud-shrouded rocky surface of Venus, and the mysterious icy satellites of Jupiter and Saturn. These views, combined with a new perception of Earth as a planet, have changed our perspective about our place in the solar system. The planets are now much more familiar to us. In many ways, Earth is no longer unique; it is one member of a larger family—the solar system. Just as our philosophical perspective has changed, so has our scientific perspective been revolutionized. The information contained in the spacecraft images has changed the planets in our perception from largely astronomical objects to geological objects. Mapping and study of their surface features allows us to think about the planets in geological terms and to make comparisons with the geology of Earth. Thanks to the results of space astronomy, we can undertake comparative planetology and describe the major themes and stages in the formation and evolution of similar bodies in the solar system.

How is this transformation accomplished? How can the images be converted into hard scientific information? What kinds of questions do geologists ask when they view planetary surfaces?

In the broadest sense, geologists want to know how a planet "works"—to discover those forces that form a single planet and influence its evolution.

The approach to determining how planets work is both simple and complex. First, one simply observes the process. However, consider the problem of observing the Earth. Only vegetation, soil, rocks, and a very tiny percentage of the upper surface are visible. The simple challenge to "observe how the Earth works" becomes a task that appears chaotic, indirect, complex, incomplete, and unconstrained. Most scientists prefer to study a more controlled and predictable environment where specific variables can be isolated, measured, and analyzed, as in the mixing of known chemical substances under strict laboratory conditions. But the fundamental issue remains: How does our planet work? We must strive to answer this seemingly overwhelming question under the less-than-perfect conditions of the real world.

Geologists begin this task by identifying and understanding the geological processes that shape and modify the surface of a planet. For example, the presence of volcanism indicates melting and redistribution of material from the Earth's interior. Earthquakes, faulting, and mountain building indicate tectonism and deformation of the crust. Streams, rivers, glaciers, sand dunes, soils, and rock debris are evidence of erosion as the atmosphere, the cryosphere, and the hydrosphere interact with and modify the solid rock surface. But there are two problems in identifying and understanding geological processes: scale and time. Many of the processes we study operate on such a grand scale (the Mississippi River, for instance) that it is difficult to observe, measure, and reduce these observations to a set of equations that characterize the process. The river system is too open, too large; there are too many variables. Time poses comparable problems. Consider mountain building and tectonics. Mountains are formed over millions of years. A devastating earthquake along the San Andreas Fault is just an insignificant movement in the whole process. The same holds true for volcanism; the eruption of Mount St. Helens was just one more volcanic event in the northwest United States, where volcanos have been building up over millions of years. Thus, although observations of now-active geological processes can provide important data, we must turn elsewhere to complete our view.

Geologists attempt to integrate processes over time and space by studying the rock record, the clues to geological history recorded in the Earth's crust. The nature and the configuration of rocks and their associated structures are traditionally defined and characterized by geological mapping. The rocks in a small portion of the crust are mapped in detail by geologists traversing the countryside. Rock composition, orientation, and structure is noted, and a map is compiled. Many such maps are pieced together until the patterns of regional geological processes are recognized (the Appalachian Mountains or Pleistocene glacial deposits, for example). More maps reveal the nature of continents and ocean basins. It was no coincidence that the global theory of plate tectonics became well stated and accepted only after the morphology and structure of the sea floors were revealed.

Yet this considers only the surface of the planet. If we are interested in what really makes a planet work, we must know what is occurring in the interior, where virtually all the mass and heat reside. Indeed, planets are much like watches. We can see the hands move or the numbers dissolve and reappear, but the mechanism for these changes, the internal workings of the watch, are hidden from our view. Unfortunately, we cannot take planets apart to see what makes them tick. We must rely on indirect evidence. For example, we can infer something about the internal structure of a planet by observing the behavior of seismic waves passing through its interior. We can also learn a lot about the interior from looking at the configuration of rocks and their surface structure. For example, the eruption of a major volcano on Earth, such as Mount St. Helens, hints that material is being melted by a thermal anomaly in the interior and then being redistributed to the surface. Faults and folded mountain belts indicate vertical and lateral movement of the Earth's crust and outer rigid layer (lithosphere) in response to stresses generated by internal activity. Thus, the global configuration of faults and folds defines global patterns of internal activity. Sea-floor spreading, continental drift, and plate tectonics provide abundant evidence for mantle convection and the mechanism by which the Earth rids itself of heat generated in its interior.

Armed with knowledge of surface geological processes, the rock record, and the nature of the interior, we can pose the next major question: What is the history of the Earth? What processes have operated in the past? Has their relative importance changed? What has been the flow of events with time? The rock record, despite its incompleteness, is our only source of information about the history of the Earth. Although all the planets are the same age, having formed about 4.5 billion years ago, two-thirds of the surface of the Earth formed in only the last 200 million years of planetary history. Thus, most of the present surface—the ocean basins—formed in the last 5 percent of planetary history. Knowledge of the other 95 percent of the Earth's history, including its early formative years, must be gleaned from a decreasingly abundant and increasingly incomplete rock record. To complete the puzzle of planetary history—in a sense, to find the missing pieces—we must look elsewhere.

Fortunately, the view of the planets provided by space exploration allows us to put the Earth in perspective and, through comparative planetology, to answer several questions fundamental to understanding

Table 2.1
Basic characteristics of inner planetary bodies.

	Mass (Earth masses)	Mean radius (kilometers)	Density (grams per cubic centimeter)
Mercury	0.055	2,439	5.43
Venus	0.815	6,051	5.24
Earth	1.000	6,371	5.52
Mars	0.107	3,390	3.93
Moon	0.012	1,738	3.34

how it works: What are the major factors that determine the nature and evolution of planets? What role does planetary size play? Are starting conditions and chemical composition the major factors? What is the role of heat generated and lost by a planet? And how does the position of a planet in the solar system affect its evolution? To address these basic questions of the nature and evolution of the planets, we will proceed by examining the gross physical characteristics, surface morphology, and structure of each of the inner planets, including our own.

Earth

Earth is the largest and densest planet in the inner solar system. (See table 2.1.) A view of Earth from space emphasizes the complex nature of the near-surface terrestrial environment: swirling clouds of water vapor within an atmosphere rich in nitrogen and oxygen, a hydrosphere dominated by oceans covering over two-thirds of the surface, polar ice caps (the cryosphere), and a significant but varied biosphere. All these interact with each other and with the solid rock surface of the planet in significant ways. However, our main concern here is the nature of the solid surface, its composition and structure. If we strip the atmosphere, the hydrosphere, and the cryosphere from the Earth, two major subdivisions or "provinces" of the crust are observed: the continents and the ocean basins. Continents are characterized by high-standing topography composed of metamorphic, igneous, and sedimentary rocks. The topographically low ocean basins are dominated by igneous rock of basaltic composition and of higher density.

Decades of seismological study have revealed the three-dimensional nature of these provinces and the structure of the planet's interior. Earth can be subdivided into a series of concentric layers distinguished by compositional differences. The outer layer, or crust, contains continents (averaging 35 kilometers thick) and oceanic areas (averaging 5 kilometers thick). The Mohorovicic Discontinuity separates the crust from the underlying mantle, a layer that is rich in iron, magnesium, and silicates and extends down to the core. Earth's dense, iron-rich core, approximately the diameter of Mars, makes up about half the radius of the planet and is composed of an outer molten zone and an inner solid zone. Earth is also characterized by mechanical layers. At a depth of approximately 100 kilometers, the temperature and pressure increases are such that partial melting of mantle rocks occurs. This partial melting changes the physical properties of the rocks, making them less rigid and more plastic. This is the basis for distinguishing an outer rigid layer (the lithosphere) from a deeper, more deformable zone (the asthenosphere). The nature, thickness, and continuity of the lithosphere will be an important theme throughout our examination of the planets.

Even a quick overview of the global geology of the Earth reveals a fundamental clue to its nature. Geological features and structures are neither evenly nor randomly distributed around the planet. The continents and the ocean basins are fundamentally different. Continents contain vast ancient shields of deformed igneous and metamorphic rocks overlain by younger sediments deposited by advancing and retreating seas. Occasionally, giant graben (vertical displacements of large crustal blocks) cross continental areas, bearing testimony to disrupting tensional forces. (The Rhine Graben and the East African Rift are examples.) Folded mountain belts, indicating compressional forces, are often aligned parallel to continental edges, with younger mountains such as the Andes being associated with active volcanism and earthquakes while older mountains such as the Appalachians are more quiescent. Occasional vast plateaus, for example the Himalayas, dominate continental topography. The ocean basins have a totally different structure. Linear midocean ridges and rises are cut by transform faults. The rocks on the ocean floor are everywhere young (less than 200 million years old). The rocks are youngest at the crests of the rises and ridges and become progressively older at greater distances. There is no evidence of the compressional tectonics that characterize the edges of continents. Rather, the areas adjacent to young continental folded mountain belts are dominated by deep linear trenches. Seamounts and volcanos dot the surface of the ocean floor, sometimes in long rows such as the Hawaiian-Emperor seamount chain.

Before World War II, our knowledge of the ocean basins was extremely sketchy. Geological concepts were dominated by what we saw and studied on the continents, and thus represented only one-third

Figure 2.1
Folded sedimentary rocks in a mountain range of compressional origin (the Appalachian Mountains of Pennsylvania) as viewed by Seasat radar. The Susquehanna River is seen in the upper right. Width of view is about 80 kilometers. (NASA photograph)

of the Earth's surface. However, the accelerated exploration of the oceans after the war added to our data on the topography, the structure, the chronology, the composition, and the magnetic characteristics of the basins. By the 1960s, this information had been combined with old ideas about continental drift into a new global theory of plate tectonics. This unifying concept correlated and explained many seemingly unrelated geological processes and phenomena. In brief, the theory proposed that the apparently rigid outer layer of the Earth, the lithosphere, was actually subdivided into a series of segmented plates that moved laterally relative to one another. The plates were created at midocean ridges and rises where basaltic material welled up from the interior. And in a continuous process known as sea-floor spreading, the recently formed material divided and spread laterally as newer material was intruded along the axis of the ridge or rise. A decrease in elevations occurred away from the ridge as the new plate cooled. Thus, ridges and rises represent divergent lithospheric-plate boundaries. But if Earth maintains a constant radius, how do we account for these rapidly forming and spreading lithospheric plates? The answer lies at convergent plate boundaries. Where two oceanic plates come together, one must be subducted, or slide under, the other. At the convergent zones, one of the relatively rigid lithospheric plates flexes and descends under the other to be remelted and reincorporated into the mantle.

Continental regions fit into the picture in several ways. At present, continents exist in the upper part of the lithosphere over one-third of the Earth's surface. Continents have formed over geologic time from the accretion of low-density crustal material in a poorly understood process. Because of their low density, they stand topographically high and, in general, they ride passively on the laterally moving lithospheric plates. Divergent plate boundaries often form in or cross continental regions. Where this occurs, major graben systems are formed. As spreading continues, oceanic crust and then an oceanic basin is created; this results in continental drift. At convergent boundaries, where continental crust forms the edge of one lithospheric plate (the west coast of South America, for example), the topographically lower and denser oceanic plate is subducted; this creates a broad zone of deformation. As the oceanic plate flexes it produces a narrow linear trench, and as it subducts the upper part of the plate and the continental crust are compressed, folded, and faulted to produce a linear mountain belt characterized by abundant earthquakes. Associated melting also produces volcanos, such as the characteristic stratovolcanos of Central America and the northwest United States. Where lithospheric convergence is dominated by low-density continental crust on both plate edges, subduction is a much more complicated process and broad, uplifted, deformed plateaus are produced (for example, the Himalayan Plateau, which resulted from convergence of India and Asia).

Tectonic activity is not as dramatic within the laterally moving plates as at their boundaries because of their relatively rigid nature. However, intraplate activity includes loading and downwarping of the lithosphere (vertical tectonics) by ice and sediments. Localized sublithospheric centers of melting (hot spots) may remain stationary and cause melting of the overlying lithosphere and extensive volcanism. As the lithospheric plate moves over the hot spot, volcanos are produced and then removed, creating a linear array of volcanos such as the Hawaiian-Emperor seamount chain.

In this fashion, the theory of plate tectonics describes the behavior of the Earth's relatively rigid and cool lithosphere (the outer layer of rocks, which is approximately 100 kilometers thick). The plates are segments of the lithosphere that move laterally relative to one another at rates measured in centimeters per year. Plates are created at divergent plate boundaries, move laterally through sea-floor spreading, and are highly modified or subducted and destroyed at convergent plate boundaries. Over geologic time, the lighter continents formed and moved laterally with the plates, constantly breaking up, colliding, and rearranging their global patterns.

Plate tectonics also provides an explanation for the vast majority of Earth's volcanic and tectonic activity. The driving mechanism for the laterally moving plates appears to be thermal convection in the mantle below the lithosphere. As the Earth attempts to rid itself of internally generated heat, the lithosphere serves as a thin thermal boundary layer between the interior and space. At diverging plate boundaries, hot material ascends to near the surface. As it spreads laterally, the new material cools, becomes denser and less buoyant, and descends back into the interior as a cooler slab, thus completing a thermal convection cycle.

One of the exciting aspects of plate tectonics is that the rate of formation and movement of the plates is so great in terms of geologic time that it can be studied as an active geological process. Current satellite tracking techniques may soon allow us to measure this motion of the continents directly. However, this tectonic activity places serious limits on understanding the Earth's history through its geological record.

The ocean basins, which comprise two-thirds of the surface, are less than 200 million years old. The continents contain older rocks, but even these are dominated on the surface by rocks less than 600

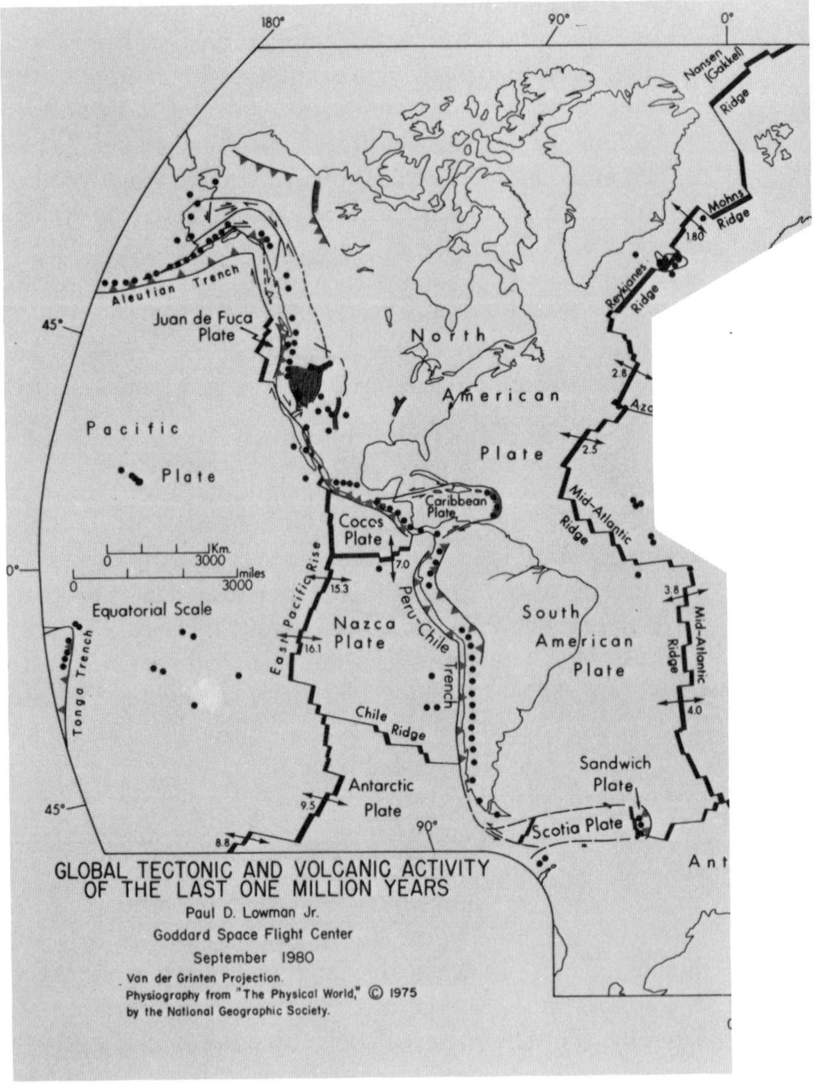

Figure 2.2
Tectonic map showing major plate boundaries and fault zones. Calculated spreading rates are indicated by arrows on oceanic ridges. Volcanos active within the past million years are represented by solid circles. (Courtesy of Paul Lowman)

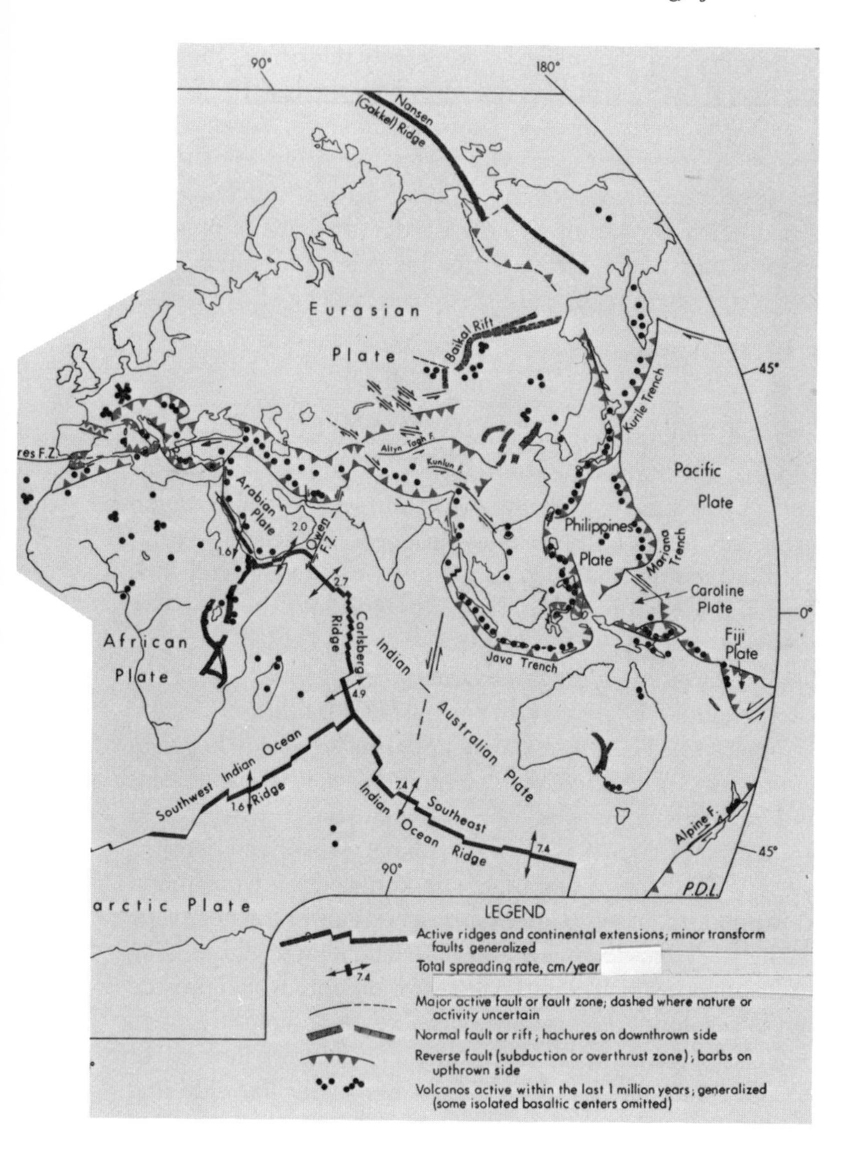

90°
180°
Nansen (Gakkel) Ridge
Eurasian
Plate
Baikal Rift
Altyn Tagh F.
Kunlun F.
Kurile Trench
Pacific
Plate
Philippines
Plate
Mariana Trench
Caroline Plate
Fiji Plate
Java Trench
Arabian Plate
Owen F.Z.
2.0
1.6
2.7
Carlsberg Ridge
4.9
African
Plate
Indian
Australian Plate
Southwest Indian Ocean Ridge
1.6
7.4
Southeast Indian Ocean Ridge
7.4
Alpine F.
P.D.L.
45°
0°
45°
90°
arctic Plate
LEGEND
Active ridges and continental extensions; minor transform faults generalized
Total spreading rate, cm/year
7.4
Major active fault or fault zone; dashed where nature or activity uncertain
Normal fault or rift; hachures on downthrown side
Reverse fault (subduction or overthrust zone); barbs on upthrown side
Volcanos active within the last 1 million years; generalized (some isolated basaltic centers omitted)

million years old. Because the surface is constantly reforming, the rock record is dominated by material produced in only the last 10 percent of the Earth's 4.5-billion-year history. There are no known rocks dating from the first 10 percent of the Earth's history, and our knowledge of the next 70–80 percent is sketchy at best. So we find ourselves with a reasonable understanding of how Earth works at the present, but a very poor understanding of its formative years and middle age. Ironically, we have a better, or more complete, history of the moon.

The moon

Earth's natural satellite has long been known to be smaller than Earth (about one-fourth the radius), to be less dense, and to lack an atmosphere and a hydrosphere. Since the moon is in synchronous rotation with the Earth, only about half the lunar surface is visible from Earth; the nature of the far side was not known until Soviet spacecraft flew past the moon and looked back. Simple observations of the near side from Earth reveal that its surface also can be subdivided into two major provinces: dark, low-lying plains (maria) arrayed in circular or irregular patches (about 17 percent of the surface) and bright, rough, highland terrain (terra). The immediately apparent similarity between these lunar provinces and the Earth's continents and ocean basins raises a fundamentally important question: Are these provinces of similar origin, despite major differences in planetary size and density?

Before the era of space exploration our knowledge of the moon was derived from Earth-based telescopic observations of the near side. In the late 1950s the United States and the Soviet Union began sending spacecraft to the moon, which over the next two decades would provide data destined to revolutionize our view of the moon and of the Earth's place in the solar system. Lunik and Ranger spacecraft flew by or crashed into the lunar surface, providing views of the far side and details of the surface at several meters resolution. Early Luna and Surveyor spacecraft soft-landed on the lunar surface, providing detailed views of a few areas. Later Luna missions sent automated rovers across the surface and returned samples to Earth. Lunar Orbiter and Zond spacecraft provided detailed orbital images of the whole lunar surface. The zenith of lunar exploration was reached with the Apollo missions, when scientifically trained astronauts explored the surface on foot and in vehicles, deployed complex and long-lasting geophysical stations, collected samples to be returned to Earth for analysis, and carried out a wide-ranging series of geological, geochemical, and geophysical observations and experiments from lunar orbit.

What types of geological processes were revealed by these data? The most striking and enigmatic features on the lunar surface were craters, which ranged in size from tiny pits on individual rocks up to giant basins more than 1,000 kilometers across. Study of telescopic images had fueled a controversy over the origin of the craters long before spacecraft exploration began, with one side favoring volcanos and the other side arguing for the impact of meteoroids. Exploration of the moon showed that the vast majority of lunar craters were indeed the result of hypervelocity impact, a process regarded as unimportant on Earth. When a projectile hits the lunar surface at speeds measured in kilometers per second, the kinetic energy produces several types of effects. Most of the energy is expended in the deformation and ejection of materials from the area below the impact point. This produces a craterlike cavity and patterns of secondary ejecta surrounding the depression. A small portion of the energy causes melting of the target rock, so the cavity is often lined with melt deposits. The transient cavity often collapses, forming scalloped or terraced walls, central uplifts, and shallower (often flat) crater floors. These effects can be truly gargantuan. The Orientale basin (figure 2.5) is approximately 900 kilometers in diameter, was originally several tens of kilometers deep, and has spread ejecta over almost an entire lunar hemisphere. All this was accomplished in minutes or, at most, hours—an extremely short period of time in terms of geological processes. No area on the moon has escaped impact cratering, and the lunar highlands are dominated by these features on all scales. Why is the moon so dominated by this process, and why is the Earth not? Clearly the lack of an atmosphere on the moon cannot alone explain the difference. We will return to this question when we discuss lunar history.

Earth-based telescopic views of the lunar maria showed lobate scarps that appeared to be analogous to the flow fronts common on terrestrial lava flows. Exploration revealed that the maria were composed of a series of lava flows emplaced over many hundreds of millions of years. However, the mode of emplacement and the associated structures were very different from those found in terrestrial ocean basins. Rather than form at a divergent plate boundary and then move laterally, lunar lavas tended to emerge in the vicinity of the large impact basins (particularly along the major ring fractures) and to flood the low-lying basin interiors. Thus, most lunar maria were found to be more comparable to terrestrial flood basalts, which emerge at very high eruption rates along major crustal fractures, than to the slower but more continuous emplacement of lavas along oceanic spreading centers.

Figure 2.3
Two views of the moon's near side as seen by telescopes from Earth. (Left) Full moon, with illumination from behind the viewer; topography is subdued and albedo enhanced. (Right) A mosaic of two half-moons, with topography enhanced and albedo subdued. (Courtesy of Lick Observatory)

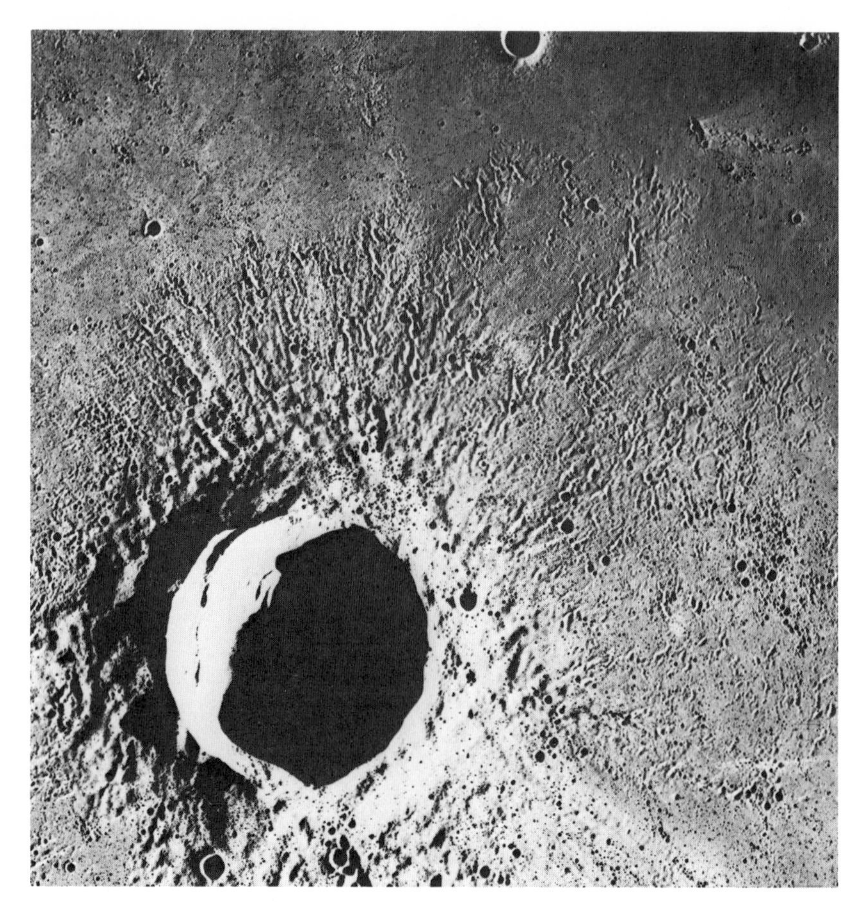

Figure 2.4
The lunar crater Euler, about 26 kilometers in diameter and 3 kilometers deep. Note raised rim and secondary craters radiating away from central crater. (NASA photograph)

Figure 2.5
The lunar Orientale basin. Note the multiple concentric rings, the outermost of which is 900 kilometers in diameter. Compare the size of the basin with the curvature of the moon and with the size of the state of Rhode Island (superposed). (NASA photograph)

Figure 2.6
Lava flows in lunar Mare Imbrium. Flows are 10–20 kilometers in width and extend across hundreds of kilometers of the mare surface. (NASA photograph)

What about tectonics and mountain building on the moon? The major lunar mountain ranges were not formed by the collision of lithospheric plates, as on Earth, but rather were uplifted almost instantaneously during the formation of the major impact basins early in lunar history. Tectonic features do exist on the moon, but some styles of tectonic features are lacking. The large number of circular craters makes detection of lateral offset on faults (strike-slip movement) easy to detect, yet there is no evidence for such movement on the lunar surface. Extensional structures (graben) are seen surrounding some of the lunar maria, and compressional features (mare arches and ridges) are seen primarily in the interiors of the maria. The distribution of these features is in no way similar to the structure of the ocean basins and to the process of plate tectonics. Also, the level of quake activity, as measured by Apollo seismometers, is much lower than that on Earth. Thus, the initial similarities between provinces on the Earth and the moon, and the possibility that plate tectonics might exist on the moon, have given way to the probability that the two planetary bodies operate in very different ways.

The nature of the lunar interior, as revealed by Apollo seismic data and returned samples, provides clues to these differences. The interior of the moon, like that of the Earth, appears to be both chemically and mechanically zoned. The lunar crust averages approximately 70 kilometers in thickness, is composed of feldspar-rich crustal material, and is global in extent. The lowlands, although basaltic in nature, are superposed on the global feldspar-rich crust, in contrast with the configuration of continental and oceanic crust on Earth. The lunar crust overlies a more iron-rich mantle and a possible central core. Seismic data show that, at present, the lunar lithosphere is about 1,000 kilometers thick, or almost two-thirds the radius of the planet. Earth's lithosphere, relative to the planet's radius, is much thinner. Analysis of Apollo samples and other data suggest, however, that the lunar lithosphere was once much thinner.

What does the geological record of the moon tell us about its history? One of the most exciting results of the Apollo program was the discovery that even the youngest mare lavas on the moon were almost 3 billion years old and thus dated from the first half of the solar system's history. The oldest lunar rocks were over 4 billion years old and thus recorded the earliest stages of planetary evolution.

The moon was formed by an accretion of debris. During the last stages of this accretion, the intense bombardment probably melted the outer several hundred kilometers of the lunar surface to form a "magma ocean." The lunar crust was formed by flotation of crystallizing feldspar, leaving an underlying residuum of more iron-rich material. As this global crust formed, so did a globally continuous lithospheric shell—a single lithospheric plate rather than multiple, laterally moving plates as on Earth. Crustal formation was complete and the lithosphere stabilized within the first several hundred million years of lunar history. The impact rate remained relatively high until sometime about 3.9 billion years ago, and the newly formed crust and lithosphere were heavily modified by crater and basin formation. The last major lunar basin, Orientale, formed about 3.8 billion years ago.

Slightly earlier, basaltic lavas had begun pouring onto the lunar surface to form the mare deposits. Lavas filled low-lying areas (particularly the major impact basins) for about 1.5 billion years, decreasing in volume until the last lavas flowed out onto the lunar surface about 2.5 billion years ago. The majority of the tectonic features formed

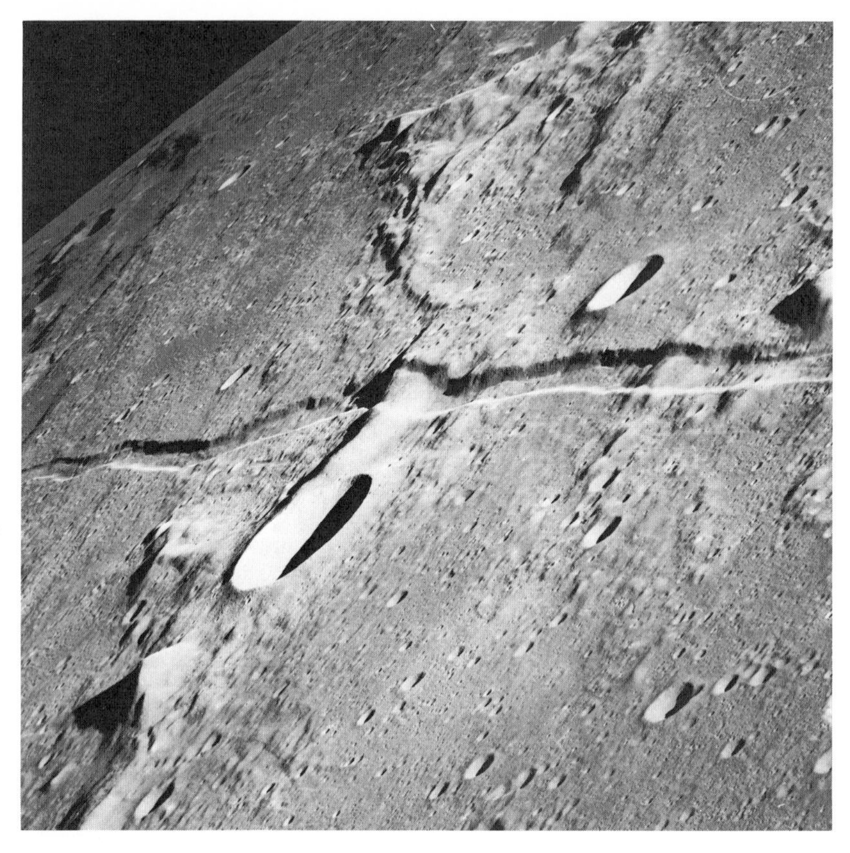

Figure 2.7
A lunar linear rille, a tectonic feature or graben of extensional origin. The graben (Rima Ariadaeus), over a kilometer in width and several hundred meters deep, formed when tensional forces fractured the crust and caused the central region to drop. (NASA photograph)

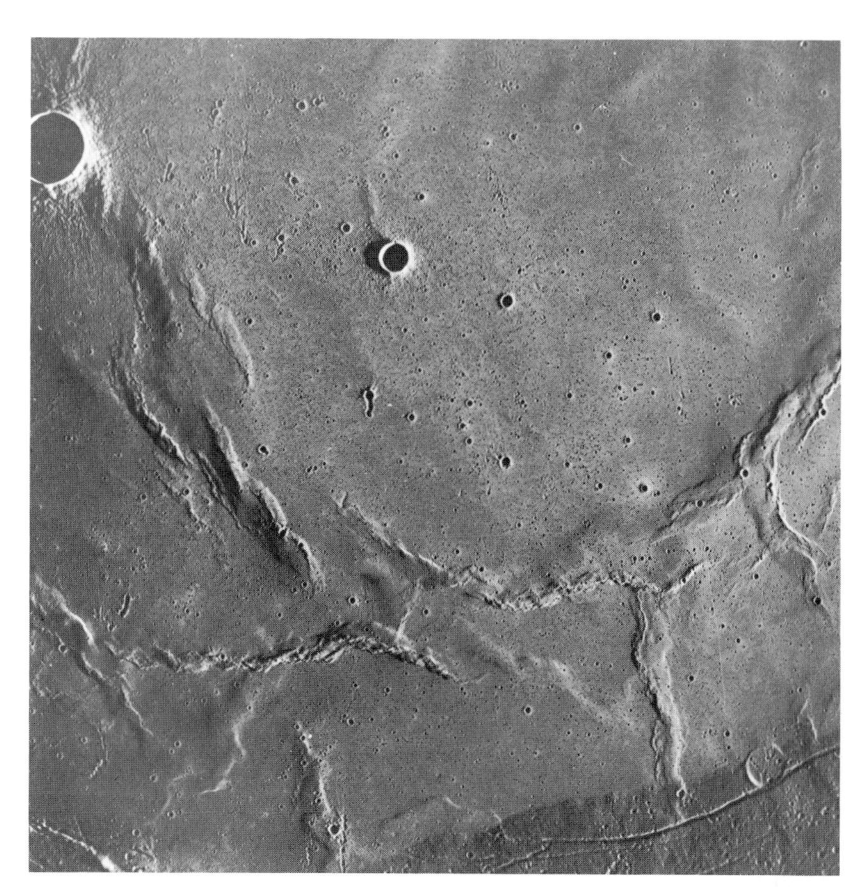

Figure 2.8
Lunar mare ridges in southern Mare Serenitatis. The somewhat sinuous arches and ridges, arrayed generally concentrically to the basin, formed when the basin subsided and underwent compression, which caused buckling and deformation. Width of image is about 150 kilometers. (NASA photograph)

during the time of mare emplacement. As the lavas flowed out onto the lunar surface and accumulated in low-lying areas, they formed a load on the lunar lithosphere. The lithosphere, in response to this vertical load, flexed and downwarped toward the basin interior, causing fractures and faults that are observed today. During this time, the moon was cooling and the lithosphere was increasing in thickness. By the end of mare emplacement, the lithosphere was thick enough to support the last of the mare load and a positive gravity anomaly, the mascons, remained. By 2.5 billion years ago, the moon had cooled sufficiently so that virtually no surface activity (volcanism, tectonism) took place. During the last half of the solar system's history, the moon has had an extremely stable lithosphere and surface, occasionally modified by impact events.

Thus the moon's initially apparent similarities with the Earth have given way to fundamental differences. Still, the moon provides a complement to the Earth in two ways. First, it provides us with a complement in time, filling in many of the missing pages of Earth's history and showing the potential significance of impact cratering as a process in early planetary history. Second, it provides us with a complement in tectonic style and thermal evolution. Unlike Earth, the moon has a single lithospheric plate, which has been stable over its entire history. Moreover, lunar tectonics are vertical, rather than lateral, and the moon rids itself of heat through conduction rather than plate recycling. What is the cause of these fundamental differences? Perhaps the answer lies in the nature of Mercury, Mars, and Venus.

Mercury

The proximity of Mercury to the sun means that it is very difficult to observe telescopically. Our knowledge before spacecraft exploration was limited to information on its size, density, broad surface composition, and orbital characteristics. Mercury is intermediate in size between the moon and Mars, yet its density is very high—much closer to that of Earth (table 2.1). On the basis of our knowledge so far, what might the planet look like? Because of its size it may have a surface like that of the moon, but the Earthlike density could produce an entirely different interior and surface evolution.

In 1974, the Mariner 10 spacecraft flew by Mercury and returned the first images of its surface. In this and two subsequent encounters, about one-third of Mercury's surface was revealed and its general moonlike appearance was confirmed. Craters of all sizes were visible on the surface. The general morphology of these craters and basins was similar to that observed on the moon. However, interesting variations were seen in the distribution of ejecta deposits, with Mercurian ejecta bunched closer to the rim of the crater than those of similar-size lunar craters. This was attributed to the higher gravity on Mercury, which inhibited the radial distribution of ballistic ejecta. Other differences in the interior structure of craters appear to be due to the influence of the physical properties of the target material. There are also a number of impact basins on Mercury, the most impressive of which is the 1,300-kilometer-diameter Caloris basin (figure 2.10). The presence of these craters and basins strengthens the possibility that impact cratering is an important process throughout the inner solar system.

Another similarity between Mercury and the moon is the presence of plains deposits. Mercury's smooth plains are similar to the lunar maria in crater density, smoothness, and the presence of mare ridges. The intercrater plains of Mercury are smooth to rolling, have a higher crater density, and occur between and around large craters and clusters of craters. The lunar maria are of volcanic origin, of course, and appear darker than the highlands. On Mercury, however, the albedo of the plains is similar to that of the surrounding cratered terrain. As a result, the origin of Mercury's smooth plains is controversial. Most researchers favor a volcanic origin, but others suggest an origin related to impact ejecta, as is known to be the case for many lunar light plains. Without additional data, the origin of the Mercurian plains cannot be firmly established.

Are tectonic features on Mercury comparable to those on the moon? Unlike the moon, Mercury is characterized by a widespread distribution of lobate scarps, which are found on virtually all types of topography and which superficially resemble lunar mare ridges. Hundreds of kilometers long and sometimes over a kilometer in elevation, these scarps appear to be compressional, in that where they intersect preexisting craters the craters have been foreshortened. The measured foreshortening has been converted to a value for a decrease in the planetary surface area equivalent to a 1–2-kilometer decrease in the radius of Mercury. Two mechanisms for the formation of the scarps have been suggested: global contraction related to the general thermal evolution of the planet, and tidal deceleration and accompanying stresses related to Mercury's tidally evolved spin-orbit resonance. Although tidal deceleration may have played a role in scarp formation, a superposed global compressive stress system due to planetary contraction seems required to account for the observed patterns. Tectonic features of extensional origin, such as the lunar graben, are rare on Mercury,

Figure 2.9
Mariner 10's approach view of Mercury. (NASA photograph)

Figure 2.10
The Caloris impact basin, about 1,300 kilometers in diameter, on Mercury. Note radial ejecta patterns and smooth plains outside basin. (NASA photograph)

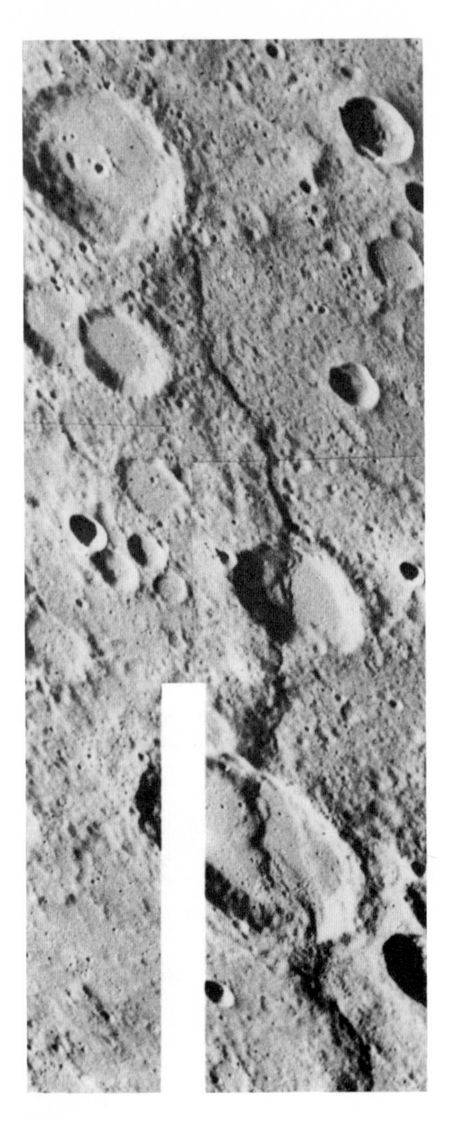

Figure 2.11
Discovery Scarp, one of the most prominent lobate scarps on Mercury. The scarp is interpreted to be a thrust or reverse fault and to indicate compression. Two large craters, 35 and 55 kilometers in diameter, are cut and foreshortened by the scarp. (NASA photograph)

and there is no evidence for major disruption and recycling of a series of lithospheric plates. Rather, Mercury, like the moon, is characterized by a single global lithospheric plate. A major difference on Mercury is that the global plate has been broken (resulting in the lobate scarps) but has not been subducted or recycled.

There are no direct seismic data concerning the structure of Mercury's interior. Mercury's very high density suggests that it is composed of 60–70 weight-percent metals and 30 weight-percent silicates. If the metal is predominantly iron and concentrated in the interior, then the core makes up about 75 percent of Mercury's radius—a radius approximately comparable to that of the entire moon.

No samples have been brought back from Mercury, and no absolute ages are known, so the interpretation of Mercurian history must rely on the chronology of superposed craters and, in turn, on assumptions about cratering rates. If the cratering rate at Mercury is generally comparable to that at the moon, then the evolution of the two planets has been similar: An initial period of heavy bombardment shaped the surface of newly formed and globally continuous crust and lithosphere. The major phase of impact crater and basin formation was followed by plains formation, including the emplacement of smooth plains of presumed volcanic origin. Scarp formation extended throughout and subsequent to the emplacement of smooth plains, but by the end of the first half of the solar system's history surface volcanic and tectonic history had virtually ceased. As with the moon, Mercury's surface has remained quiet and stable for the last half of the solar system's history.

Mars

Mars is approximately half the diameter of Earth, and yet its density is more comparable to the moon's. Unlike the moon and Mercury, Mars has an atmosphere, which is dominated by carbon dioxide. At present temperatures and pressures, liquid water is not stable on the surface of Mars. The presence of an atmosphere, in combination with polar ice caps and an inclined rotational axis, produces "seasons," as evidenced by long-standing observations of seasonal and yearly regional changes. These same changes have made the possibility of life on Mars the subject of much speculation. Information on the surface morphological features of the planet was lacking before spacecraft missions, however.

The first spacecraft to encounter Mars, Mariner 4, flew by the planet in 1965 and obtained 22 images, primarily from the southern hemisphere. The density of craters revealed by these images was a major disappointment, particularly to those who had anticipated a wider diversity of geological and perhaps biological processes. Mariners 6 and 7, which flew by Mars in 1969, obtained about 60 near-encounter pictures extending to the south polar region. These photos, too, revealed primarily cratered terrain, and after their analysis Mars was considered to be dominated by ancient cratered terrain, with no evidence for processes like those on Earth. Thus, it was concluded that the Martian atmosphere had never been like Earth's, and that water had not played a significant role. Mariner 9, which went into a near-polar orbit around Mars in 1971, changed this view by documenting the presence of giant canyons, channels, gigantic shield volcanos, a variety of permafrost-related features, and a wide range of wind-related features. The Viking missions, involving two orbiters and two landers, encountered Mars in 1976 and continued documenting a rich diversity of geological processes on the planet.

As suggested by early spacecraft investigation, much of the Martian surface is cratered. Mars displays the same range of crater sizes as the moon and Mercury, including basins over 1,000 kilometers across. However, examination of fresh craters (figure 2.12) indicates that the ejecta deposits of many craters differ radically from the radial ballistic-ejecta patterns on the moon and Mercury. Many Martian craters are dominated by lobate ejecta petals and ramparts, giving the impression of an impact into a partly fluid, water- or ice-rich substrate. These craters hint at a vast reservoir of near-surface volatiles and perhaps an ancient atmosphere that could support liquid water.

Plains of volcanic origin appear to dominate Mars's northern hemisphere and also occur in the large impact basins and intercrater areas of much of the heavily cratered terrain. Many examples of lobate flow fronts and volcanic constructs are known (shields, domes, cones, fissures). As on the moon and Mercury, volcanism appears to have greatly modified the more ancient cratered terrain. The most spectacular examples of volcanic terrain on Mars are the major shield volcanos, most of which are in the Tharsis region. Rising over 20 kilometers from the surrounding plains, these massive accumulations of lava are hundreds of kilometers across and dwarf the island of Hawaii and even Mount Everest. In general, Mars exceeds the moon and Mercury in diversity of volcanic land forms and in volume of lava per unit area.

What is the nature of tectonic processes on Mars? There is no evidence for segmentation and lateral redistribution of lithospheric plates analogous to plate tectonics on Earth, and there is no evidence of the global scarps seen on Mercury. However, lunarlike linear rilles

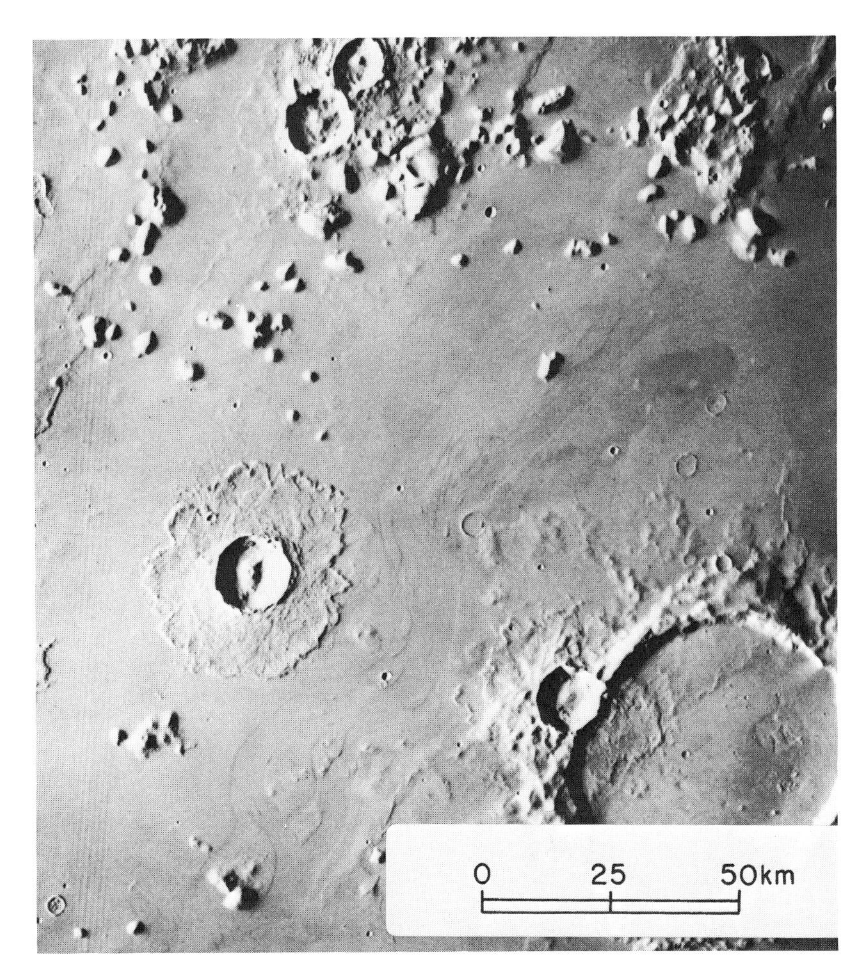

Figure 2.12
A Martian crater (about 15 kilometers in diameter) with a fluidized ejecta pattern suggesting the incorporation of subsurface volatiles into the ejecta deposit. A channel pattern can be seen meandering past the crater to the east. (NASA photograph)

Figure 2.13
The gigantic Martian shield volcano Olympus Mons as viewed from the south by the Viking orbiter. The volcano rises some 25 kilometers from the surrounding plain and is about 600 kilometers across. The complex summit caldera is 90 kilometers across. (NASA photograph)

(graben) and mare ridges abound. These occur around and within filled impact basins, as on the moon, but also around large volcanos and around the unique Tharsis region. Tharsis is several thousand kilometers across and rises up to 10 kilometers above the surrounding terrain. Capped by numerous major shield volcanos, it is the locus of linear rille and mare ridge development as well as the center for prolonged and extensive volcanic activity. A major fault valley, Valles Marineris (figure 2.14), extends several thousand kilometers radially from the center of Tharsis and is hundreds of kilometers across and several kilometers deep. The traditional explanation for Tharsis is that mantle activity caused uplift and fracturing of a broad dome, accompanied by extensive volcanic activity. This interpretation includes vertical tectonics as on the moon, but the emphasis is on vertical uplift rather than the passive loading and downwarping seen in the lunar mare basins. Alternative interpretations suggest that the Tharsis topography may have been largely constructed by volcanism, rather than uplifted, and that the tectonic features may have been induced by loading.

The discovery of major channel features on Mars by Mariner 9 led to a reexamination of the possible role of water in the planet's history. Images from Mariner 9 and Viking (figure 2.15) showed evidence of the previous flow of surface water in the form of large outflow channels containing central streamlined islands. Most of these channels are found along the boundary between the southern cratered terrain and the northern plains. Smaller stream patterns found in the form of channel networks, primarily in the older southern hemisphere, suggest that water may have flowed more extensively over the surface in earlier Martian history. That many such channels form in collapsed chaotic terrain suggests a sudden release of a large volume of water from layers near the surface. Both these observations strongly indicate that the atmosphere of Mars has changed over time. Perhaps Mars's atmosphere was once considerably different than at present.

The polar regions of Mars are composed of a permanent cap of water ice and a seasonal cap of carbon dioxide ice. A laminated or layered terrain, apparently composed of dust and ice, surrounds the poles and is in turn often surrounded and embayed by extensive dune fields. At slightly lower latitudes surrounding the poles are large polygons of fractured ground (figure 2.16) and other structures that suggest alternate freezing and thawing of ice-rich soil.

The two Viking landers provided closeup views of the Martian surface (figure 2.17) showing a cratered and windswept surface. Changes since the spacecraft landed in 1976 have been minimal, which suggests

Figure 2.14
A portion of the extensive Valles Marineris system on Mars. Canyon walls, initially formed by faulting, have been heavily modified by landslides. The walls are up to 3 kilometers high; the canyons are 50–75 kilometers wide. (NASA photograph)

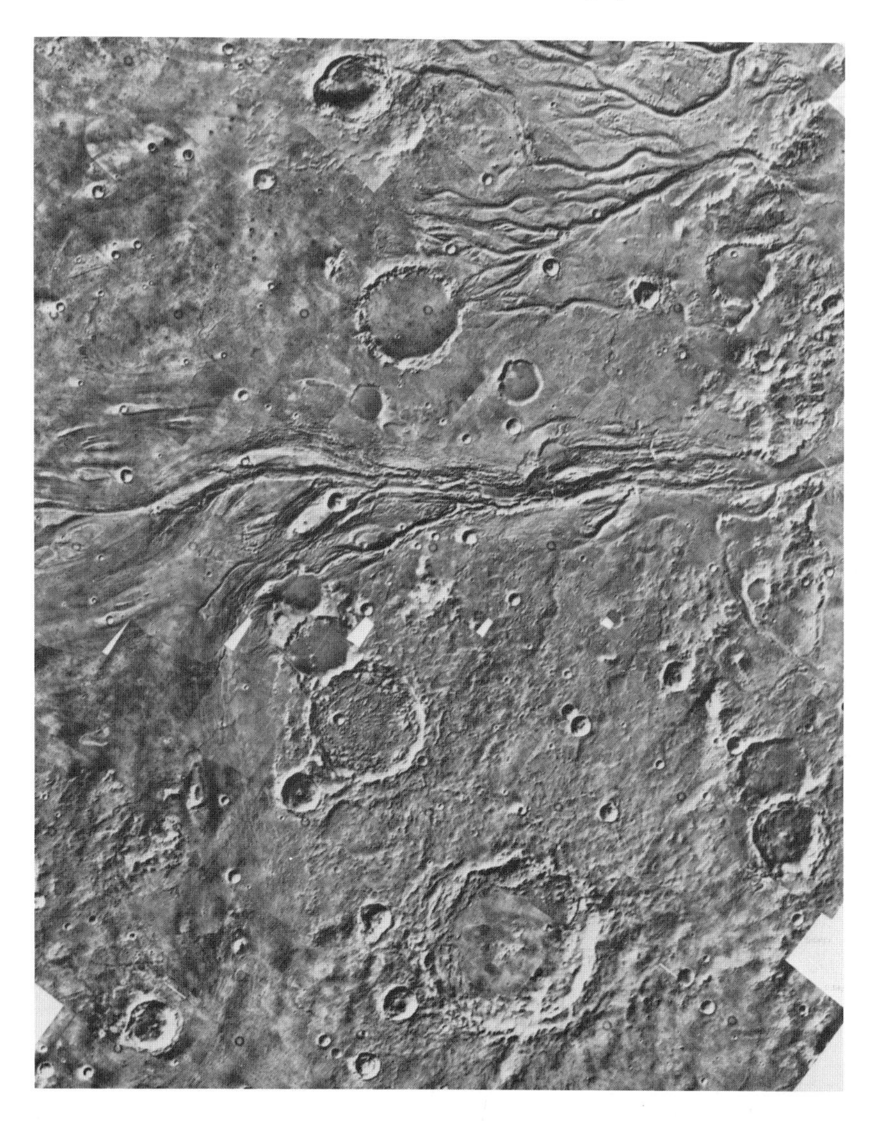

Figure 2.15
Channels in the Lunae Planum–Chryse Planitia region of Mars. Sinuous and dendritic channels extend from west to east across craters and plains strongly suggesting that flowing water was once present on the surface. The area shown is approximately 300 kilometers across. Viking 1 landed in Chryse Planitia, several hundred kilometers to the east. (NASA photograph)

Figure 2.16
Polygonally fractured ground typical of extensive areas of northern high-latitude plains on Mars. Cracks divide the terrain into irregular polygons ranging from a few to 10 kilometers in diameter. The pattern is comparable to those seen on Earth in areas where the ground undergoes alternate freezing and thawing, yet the features on Mars are about 10 times larger. (NASA photograph)

Figure 2.17
A scene from the Viking 1 landing site (Thomas A. Mutch Memorial Station) in Chryse Planitia. In the background can be seen the raised rim of an impact crater several hundred meters in diameter; in the foreground the block-littered surface has been shaped by wind erosion. (NASA photograph)

that the dune forms and surface wind-scour features are not the result of seasonal or annual wind and dust storms.

Available data suggest that Mars has a dense central core and a low-density crust that formed early in its history, and that Mars's geological history, despite some similarities to Earth's in processes, is more like that of the moon and Mercury. The history of Mars was characterized by early crust formation and by intense bombardment, which produced craters and large impact basins. The early-formed and globally continuous lithosphere shows evidence of early thickness variations, but its thickness evened out and increased with geologic time. Basaltic volcanism dominated the postcratering period of Martian history, much as it did in the lunar maria. Though most of the volcanic deposits were emplaced in the first half of the solar system's history, local volcanism continued into the last 2 billion years. Some parts of Mars have volcanic units with essentially no superposed impact craters, which suggests that local volcanic activity may still be going on. Despite the possibility of relatively recent activity and the variable nature of dust storms and polar caps, Mars's geological history still has been

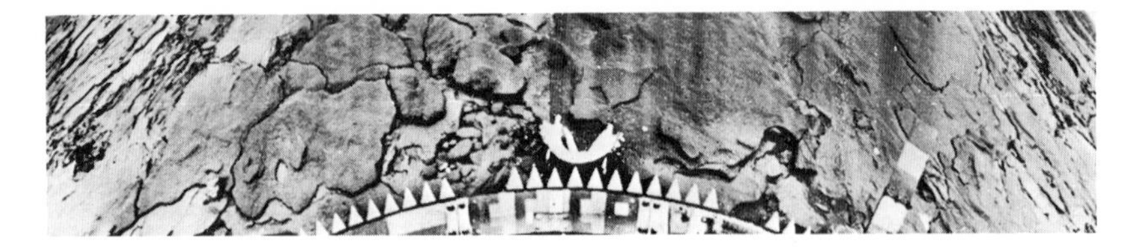

Figure 2.18
A view of the surface of Venus taken and transmitted by the Soviet Venera 14 spacecraft in March 1982. The base of the spacecraft can be seen in the foreground. Layered rock extends toward the horizon, visible in the upper left and right corners. A dark layer can be seen to overlie a lighter layer on the left side of the image.

more analogous to that of the moon and Mercury than to that of Earth.

Venus

The moon, Mercury, and Mars are all approximately half the diameter of Earth or smaller, and all have single, global, lithospheric plates that underwent heavy bombardment early in their histories, followed by extensive volcanism. These one-plate planets with ancient surfaces contrast dramatically with Earth's active, youthful, multiplated, laterally moving lithosphere. Is planetary size, then, a major factor in lithospheric fragmentation and continued evolution? Would all planets the size of Earth have adopted plate recycling as a way to deal with internally generated heat? Venus, approximately the same size and density as Earth, offers the opportunity to find out.

Although Venus has many similarities to Earth, one of the major differences is its hot, dense, carbon dioxide atmosphere, which results in surface temperatures of approximately 730° K and pressures of 95 bars. Unfortunately, this dense cloud layer precludes visual observation of surface features from Earth-based or orbiting spacecraft. Thus, investigations of Venus have had to rely on observations at radar wavelengths (where topographic and imaging data could be obtained) and on the few normal images obtained by Soviet spacecraft that descended through the atmosphere and landed. Earth-based radar observations from Arecibo, Puerto Rico, and Goldstone, California, have detected features interpreted as craters, rifts, and volcanos. But the most comprehensive data come from the Pioneer mission, which placed a spacecraft into orbit around Venus and sent multiple probes into the atmosphere. The spacecraft carried a radar altimeter that provided topographic data over more than 95 percent of the surface at a horizontal resolution of about 100 kilometers and a vertical resolution of greater than 200 meters. This basic data set provides a fundamental, but low-resolution, view of Venusian topography.

The terrain of Venus can be subdivided into lowlands (about 20 percent of the surface), rolling plains (70 percent), and highlands (about 10 percent). The highlands are concentrated in three main areas: Ishtar Terra, Aphrodite Terra, and Beta Regio. Ishtar Terra is characterized by relatively steep slopes and stands several kilometers above the mean planetary radius. The western part of Ishtar (Lakshmi Planum) is a vast plateau 2,500 kilometers in diameter, which is surrounded and cut by linear mountain ranges rising up to 3 kilometers above the plain. A massive mountain range (Maxwell Montes) occurs in Eastern Ishtar, rising 11 kilometers above mean planetary radius—higher than Mount Everest above sea level on Earth. The topography of Ishtar Terra is unlike any seen on the smaller inner planets. Recent high-resolution data obtained by Don Campbell at the Arecibo Observatory show bands 15–20 kilometers wide within and parallel to these mountain ranges; these bands are strongly reminiscent of the fold belts typical of Earth's convergent zone mountains.

Other highland areas show different characteristics. Beta Regio contains a central linear depression and several equidimensional mountains, which have been interpreted as a rift zone and volcanos, respectively. Aphrodite extends along the equator for more than 10,000 kilometers and contains numerous riftlike depressions in its central interior. The lowland areas are relatively smooth and are concentrated in roughly circular and linear areas. Midlands or rolling uplands are the most extensive terrain type and contain a diversity of topographic features, including possible rift zones, volcanos, and impact craters. Thus, our present knowledge of Venus hints strongly at a geological diversity comparable to some aspects of both Earth and the smaller inner planets, including impact cratering, volcanic activity, tectonic activity, possibly compressional deformation and folded mountain belts, and perhaps "continents." At the present radar resolution, however, we can only characterize the physiography; characterization of the geological processes responsible for this physiography must await higher resolution. We cannot yet answer the question of the possible presence of plate tectonics on Venus.

The nature of Venus's interior and its thermal evolution are also poorly known. On the basis of its mean density and the low-density, high-radioactive-element-concentration rocks measured by Soviet Venera landers, Venus has probably differentiated a crust. However,

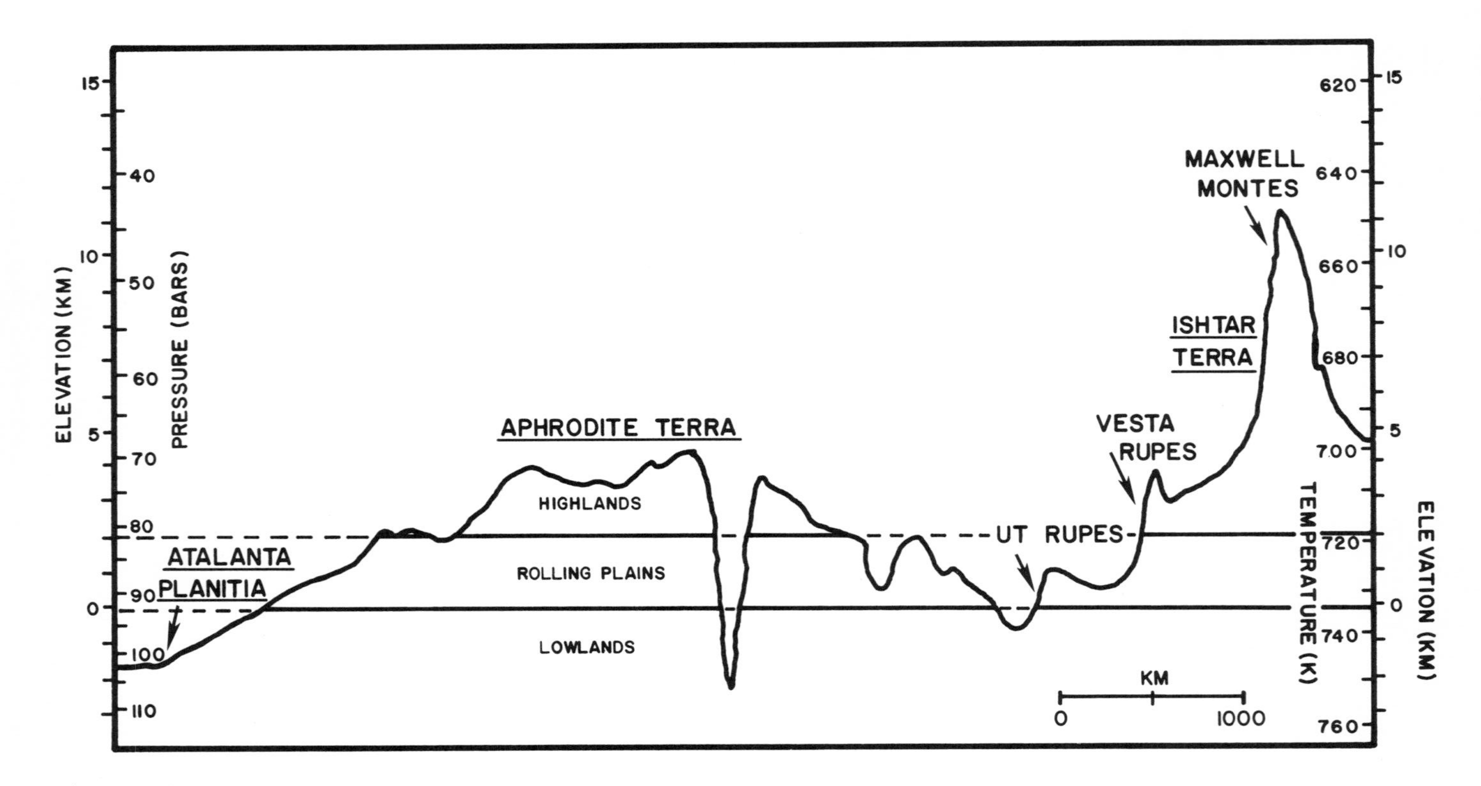

Figure 2.19
A schematic cross-sectional diagram of the topography and the physiographic provinces of Venus and their atmospheric environment.

there is no positive evidence for a central metallic core, and Venus, in contrast to Earth, has little or no magnetic field of internal origin.

Thus, much of Venus remains an enigma, and exploration of this planet offers one of the most exciting and potentially rewarding scientific challenges of the future. In addition to completing our survey of the inner planets, future exploration of Venus may tell us an immense amount about how Earthlike planets operate. More important, Venus may provide significant clues to Earth's history in its formative years—the first two-thirds of planetary history now missing from Earth's record.

Summary

What have we learned from the last 25 years of planetary exploration concerning our basic question, "How do planets work?" Exploration of the moon, Mercury, and Mars has provided us with significant, but sometimes sketchy, information on their geologic processes, their histories, and their interiors. On the basis of this information, several themes emerge which have important implications for Earth. First, these planetary bodies formed a solid crust and lithosphere very early in their history, and these features have remained stable and intact for more than 4 billion years. Thus, these single-plate planets offer, in contrast to Earth, a passive record of external and internal events over most of planetary history.

One of the most exciting revelations from this record is the influential role that impact processes have played in planetary history. The impact craters on the surfaces of the moon, Mars, and Mercury bear testimony that the impact rate was much higher in the inner solar system during the first 600–800 million years of planetary history than later on. Impact cratering and basin formation was very important in planetary heating, crust formation and modification, and creation of thermal anomalies. Although the geological record for this period on Earth is missing, we must conclude that impact cratering played at least a comparable role in early Earth history. A stage of basaltic volcanism also appears to be common to the smaller inner planetary bodies. Even before the end of heavy bombardment, radioactive heating of planetary interiors partially melted the mantle, and basaltic lavas emerged to the surface. The heavily cratered crusts were partially flooded, and this phase of volcanism was largely completed in the first half of the solar system's history. Tectonic activity on these one-plate planets was primarily vertical. Lithospheric loading and downwarping, and the possibility of doming and fracturing caused by internal

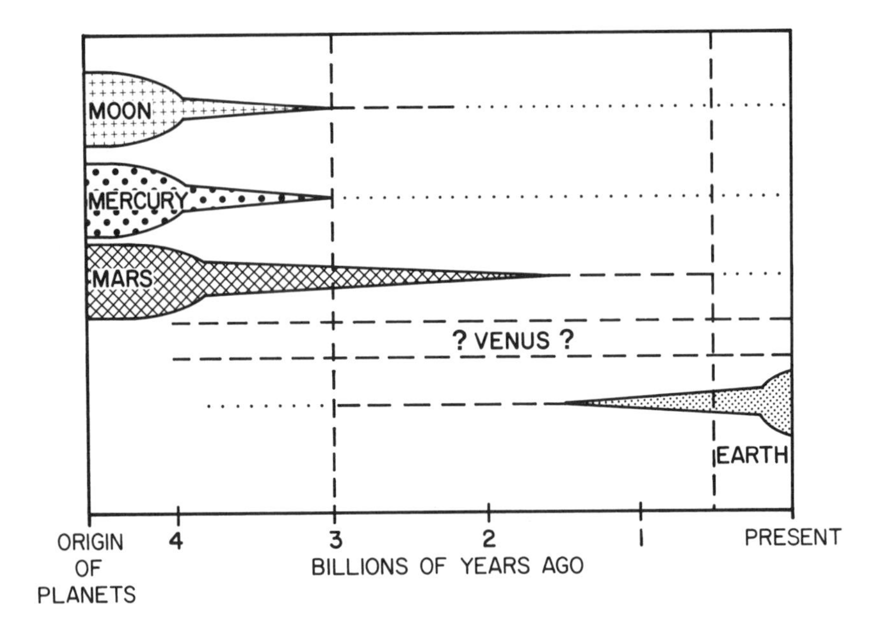

Figure 2.20
Ages of planetary surface units. The area plotted for each planet is an estimate of the percentage of the present surface area of different ages. For the moon, 80 percent of the present surface was formed in the first 600 million years of lunar history whereas less than 20 percent (mare basalts) was emplaced between 3 and 4 billion years ago. Since that time, only minor deposits, primarily impact craters, have been formed (represented here by dots). The record for Mercury appears comparable to that for the moon. The cratering flux of Mars is uncertain, and this results in uncertainties in the percentage of surface area formed between 1 and 3 billion years ago. Although volcanism on Mars appears to have extended well beyond the end of similar activity on the moon, the vast majority of exposed surface units were formed in the first half of the solar system's history. On Earth, over two-thirds of the surface, the ocean basins, was formed less than 200 million years ago, and surface rocks older than 3.5 billion years are rare. The ages of surface units on Venus are unknown; craterlike features suggest that some regions may be ancient, whereas plateaus such as Ishtar and the Maxwell Montes hint at youthful structure.

processes, dominate vertical tectonic activity. The details of tectonics and the actual state of stress in the lithosphere are related to changes in the thermal evolution of the planet. In the case of Mercury, the changes were sufficient to cause enough of a decrease in planetary radius to bring about major disruption, but not subduction, of the lithosphere. The range of tectonic activity on these bodies may provide clues to how plate tectonics began on Earth.

Several major factors could be responsible for the similarities and the differences in the inner planets. It is thought that there were distinctive and systematic temperature and pressure variations in the collapsing solar nebula at the time when the planets were formed. Certain elements and minerals could form in one part of the solar system but not in another. Thus, position in the solar system and initial chemistry must be an important factor. The high density of Mercury, for example, seems to be related to its proximity to the center of the solar system and the stability of iron there during planetary formation. Two other major factors seem to be related: planetary size and how a planet obtains and loses heat. The small inner planetary bodies appear to have cooled relatively rapidly very early in their history, so that they evolved a continuous global lithosphere. Over most of their history, then, they got rid of internal heat primarily by conducting it through the lithosphere. The larger Earth, on the other hand, gets rid of its heat primarily by continuous recycling of laterally moving lithospheric plates. Why does this fundamental difference exist? If we assume that the distribution and quantity of heat sources is approximately the same per unit volume, then a simple explanation might be that the smaller planets, with vastly greater surface areas per unit volume, could cool more rapidly and produce a thicker, more stable lithosphere early in their history.

Understanding the significance and relative importance of these factors requires additional information. The most obvious gap in our understanding of the inner planets is Venus. Exploration to date has hinted at diverse geological processes, some possibly Earthlike and others perhaps like those of the one-plate planets. High-resolution images of Venus, taken by a radar imaging system in orbit around the planet, would provide data to determine the presence or absence of plate tectonics and would allow scientists to compare the two large inner planets (Earth and Venus) and contrast their characteristics and evolution to those of the smaller bodies. Over the next 25 years, this mission, as well as the return of samples from Mars and the far side of the moon and the orbiting of spacecraft carrying geochemical, geological, and geophysical experiments around the moon and Mercury, could allow us to make major additional strides in understanding the nature and evolution of Earth's immediate neighborhood, the inner solar system.

Reading

Beatty, J. K., B. O'Leary, and A. Chaikin, eds. *The New Solar System*. Cambridge, Mass.: Sky, 1981.

Carr, M. H. *The Surface of Mars*. New Haven: Yale University Press, 1981.

Guest, J. E., and R. Greeley. *Geology on the Moon*. London: Wykeham, 1977.

Guest, J., P. Butterworth, J. Murray, and W. O'Donnell. *Planetary Geology*. New York: Wiley, 1979.

Hamblin, W. K. *The Earth's Dynamic Systems*. Minneapolis: Burgess, 1978.

Head, J. W., and S. C. Solomon. "Tectonic Evolution of the Terrestrial Planets." *Science* 213 (1981): 62–76.

Head, J. W., S. E. Yuter, and S. C. Solomon. "Topography of Venus and Earth: A Test for the Presence of Plate Tectonics." *American Scientist* 69 (1981): 614–623.

Murray, B., M. C. Malin, and R. Greeley. *Earthlike Planets: Surfaces of Mercury, Venus, Earth, Moon, Mars*. San Francisco: Freeman, 1981.

Mutch, T. A. *Geology of the Moon: A Stratigraphic View*. Princeton University Press, 1972.

Mutch, T. A., R. E. Arvidson, J. W. Head III, K. L. Jones, and R. S. Saunders. *The Geology of Mars*. Princeton University Press, 1976.

Pettengill, G. H., D. B. Campbell, and H. Masursky. "The Surface of Venus." *Scientific American* 243 (1980): 54–65.

Taylor, S. R. *Planetary Science: A Lunar Perspective*. Houston: Lunar and Planetary Institute, 1982.

3

Exploration of the Moon

John A. Wood

It has been nearly a decade and a half since the Eagle touched down on the moon. Somehow I never thought the Apollo adventure would become part of history. It was so dramatic, so bold, so difficult, so demanding of the utmost that technology had to offer that it seemed to belong permanently in the future. And, indeed, we have not attempted another feat so ambitious since.

It is common knowledge that the $40 billion Apollo program was launched by President John F. Kennedy to demonstrate U.S. superiority in the sensitive arena of space technology, not primarily to carry out scientific research. But Apollo did permit a very large amount of science to be done. Most of what we know now about our satellite has come from that short, exhilarating period from 1969 to 1972 when we flew men to the moon.

For perspective, it is interesting to recall what pre-Apollo scientific publications said about the moon. Most confined themselves to discussions of its orbit, its craters, and its apparently powdery surface. Perhaps wisely, they did not speculate about its composition. There was a general sense that the moon must be made of rocky materials similar to those of the Earth—not entirely identical to those of the Earth, however, because it was known from astronomical measurements that the mean density of the moon is much less than that of the Earth (3.34 vs. 5.52 grams per cubic centimeter).

The most obvious property of the moon, one visible by naked eye from the Earth, is its division into two terrains: the dark, smooth-looking maria ("seas") and the light-colored, rugged, cratered terrae ("continents," or "highlands"). This distinction was first made by Galileo in 1610.

The nature of the lunar maria was much discussed in the 1960s. Two prominent lunar scientists, Harold Urey and Tom Gold, had unorthodox ideas about the maria. Gold was convinced that dust particles on the lunar surface became electrostatically charged and repelled one another so strongly that the dust acted like a fluid, flowing down slopes and ponding in the low places. According to Gold, the maria were huge ponds of fluidized dust. Urey believed the moon had not had a hot igneous history and so was still composed of primitive, undifferentiated planetary material, similar in nature to the class of stony meteorites known as chondrites. At one point he proposed that the maria were the dried-up beds of watery seas that had once lapped against the lunar highlands. Moreover, he suggested that members of a certain subclass of chondritic meteorites which are rich in organic matter and bound water were, in fact, lumps of the lunar seabeds that had been knocked off the moon by the impacts of other meteoroids. Most other observers of the moon took a more prosaic view, however. To them, the dark smooth maria looked much as terrestrial flood basalt plains—lava flows—would be expected to look from a distance. Lava flows are an eminently reasonable feature to expect to find on the surface of a planet or satellite.

I was an innocent bystander to the lunar debates of this era, but like most of my colleagues engaged in meteorite research I wrote a proposal to NASA to participate in the study of lunar samples to be collected by the astronauts, if and when. The possibility that we might actually receive pieces of the moon began to seem real to me only in 1967, when those of us who had been accepted into the Apollo lunar-sample program were summoned to the new Manned Spacecraft Center in Houston. For several days, we goggled at the complex facilities being set up in the Lunar Receiving Laboratory and watched a technician demonstrate the constraints of an Apollo pressure suit and the use of special tools to sample the lunar surface. We listened intently to details about the funding of our research grants and the distribution of the hypothetical samples. One handout, entitled "Guidelines for Principal Investigators," read as follows:

> There is no guarantee that the Investigators who have been selected will receive a lunar sample. The task of returning surface materials from the moon is a challenging one and every attempt will be made to return about fifty pounds on the first Apollo mission. Thus, it may not be possible to provide samples for all of the selected Principal Investigators (PI's). Even if the total planned weight of about 50 pounds of material is returned, there may not be any suitable material for each investigation. These lunar materials are U.S. Government property. . . .

Figure 3.1
Full moon as viewed from an Apollo spacecraft above the moon's east limb. This oblique perspective is never seen from Earth, since the moon always keeps the same face pointed toward us. From this angle, part of the far side is visible (right); note the absence of maria. White arrow indicates landing site of Apollo 11 at southern edge of Mare Tranquillitatis. (NASA photograph)

When the fateful summer of 1969 arrived, my research group consisted of Ursula Marvin, John Dickey and Ben Powell (postdoctoral associates fresh from Princeton and Columbia, respectively), Janice Bower (our electron microprobe analyst), and me. We did our best to plan just what we would do with a lunar sample once we had it in our hands. We held a television party the night of the Apollo 11 moonwalk, and then waited and wondered during the weeks when the samples were being unpacked and reconnoitered and held in biological quarantine. Finally a TWX arrived from Daniel H. Anderson, curator of the Lunar Receiving Laboratory in Houston:

I have been authorized to allocate to you 10 grams of lunar material. Of this 5 grams will be material less than 1 millimeter grain size and 5 grams will be 1 millimeter to 1 centimeter.

This material will be ready for pickup on September 17, 1969.

You are a member of the "New England group" of MIN-PET P.I. consisting of Frondel CMA Simmons CMA Skinner and Wood.

Your troupe will get 22 thin and polished sections from 11 rocks. A chip of from 5 to 20 grams will accompany 9 of the rocks. This material will circulate among members of the group per instructions to follow. . . .

We and all the other Apollo investigators fell to work.

The first manned lunar landing, Apollo 11, had touched down on the Mare Tranquillitatis. A mare landing site was chosen not because of any interest NASA had in the debate about the nature of the maria, but because it was the smoothest, easiest place to land. Mission planners had worried about how successful the astronauts would be in maneuvering the descending Lunar Excursion Module to avoid craters and boulders.

The surface of the moon turned out to be covered with a meters-deep layer of loose dust and grit and rock fragments (mainly dust). This was expected from Earth-based studies of the physical properties of the lunar surface, and also from general principles: The moon has been pelted by high-velocity meteoroids since it was formed, and these must have systematically pulverized its outermost skin. The dust showed no tendency to flow as an electrostatically charged fluid, nor did it tend to vacuum-weld into a coherent mass (another possibility that had been raised). It behaved, in fact, just like dust on Earth. This seems a trivial observation, yet it illustrates what may have been the most profound lesson Apollo taught us: We had exaggerated the potential "weirdness" of other bodies in the solar system. In fact, the moon has obeyed the same laws of geological and mineralogical evo-

Figure 3.2
Southern Mare Tranquillitatis, looking west. White circle indicates Tranquility Base. The mare surface is dark and relatively smooth. At upper left, a portion of the central highlands looms pale and rugged. This photograph taken by the unmanned Lunar Orbiter V spacecraft in 1967; a flaw in the onboard processing of the film caused the wormy-looking blemishes. (NASA photograph)

lution we find on Earth, if allowance is made for the disruptive effects of meteoroid impacts, small and large, on the moon. (Earth actually also suffers meteoroid impacts, but their cratering and shattering effects are eclipsed by the far more vigorous forms of geological activity, such as erosion by running water.)

So the samples of Mare Tranquillitatis collected by Neil Armstrong and Buzz Aldrin were broken-up fragments, some small ("soil") and some larger ("rocks"). But what were they made of? Not primitive chondrites, it turned out, and not ocean deposits, but basalt—solidified lava—just as the less colorful pre-Apollo voices had predicted. The maria are covered with huge flows of flood basalt.

The Tranquility basalts and the basalts collected in other maria by later Apollo missions have been studied intensively in laboratories around the world. The lunar basalts differ from the volcanic rocks we are familiar with in some interesting ways (they are depleted in relatively volatile elements, such as sodium, and strikingly enriched in the element titanium), but still they are kissing cousins of the basalts that pour out of Kilauea in Hawaii and erupt in the depths of Earth's oceans.

Basalt is formed when the minerals of a planet's mantle are heated and begin to melt. The composition of the partial melt generated in this process is not the same as that of the "source rock." Certain elements prefer to go into the liquid; others prefer to remain in the unmelted crystals. The liquid of special composition produced under these circumstances is basaltic lava.

The exact composition of a basalt depends upon the minerals that were present in the source region (the proportions of these minerals, however, is not important); the depth and therefore the pressure exerted on the source region; and the temperature, which governs the degree of partial melting that occurred before the lava was removed from the source region by eruption to the surface.

Using special high-pressure furnaces, several research groups attempted to mimic the lunar source regions by generating melts having the same compositions as the basalts found at Tranquility Base and at several other mare sites visited by the later Apollo missions. They found, from the pressures required, that mare basalts had come from several different depths in the moon, ranging from 150 to 400 kilometers (or even more). Isotopic and trace-element studies established that the source regions had consisted not of primitive, "virgin," chondritic rock, but of rock that had already been through one or more cycles of igneous activity. This was in spite of the fact that the mare basalts are quite old. Studies of the basalts' content of certain isotopes revealed how long it had been since they had erupted on the lunar

Figure 3.3
"Mug shot" of a basalt sample from the Apollo 15 mission. Similar photographs were taken of all lunar rocks, from every perspective, before they were subdivided. Gas bubbles in this mare basalt were trapped when the lava solidified around them; however, the gas in the bubbles leaked away eons ago. (NASA photograph)

surface. For example, the isotope potassium 40 decays to argon 40, with a half-life of 1.28 billion years. Although argon is a gas, the argon 40 finds itself trapped inside crystals when it is formed—the same crystals that contain the element potassium. It is retained unless the rock is very hot, in which case the argon 40 diffuses out of its crystals and is lost. After lavas erupt they cool, and thereafter their argon 40 accumulates. If one measures their present content of potassium 40 and argon 40 and knows the half-life of potassium 40, it is relatively simple to calculate how long it must have taken to produce that much argon 40, and therefore how long it has been since the basalt erupted and cooled.

It was found that the lunar basalts show a range of ages. Some are as old as 3.9 billion years, some as young as 3.2 billion years. The solar system in general was formed 4.6 billion years ago, and there is evidence that the accretion of the planets took about 100 million years. Thus, beginning about 600 million years after the moon was largely "assembled," basaltic lavas began to pour out onto the lunar surface. The eruptions continued for another 700 million years. (This undoubtedly understates the period of lunar volcanism. It may very well have begun earlier than 3.9 billion years ago, but the record of events earlier than that time is terribly scrambled. In our six missions to the moon it is unlikely that we visited the youngest lava flow, so probably there were additional eruptions more recently than 3.2 billion years ago.)

A fairly straightforward way to look for young mare areas is by counting the impact craters visible in photographs of the maria. To make comparisons of one area with another possible, one calculates the number of craters larger than some particular size, per square kilometer, for each area. If meteoroids have been falling at a steady rate on the moon, then the older a mare surface is the greater its crater density will be. We can establish the rate at which craters appear on the mare surfaces ("calibrate the system") by counting the craters around one of the Apollo landing sites, where the astronauts collected basalt samples whose ages have subsequently been measured by radiometric techniques. Using this factor, it is possible to attach ages to all the mare surfaces whose craters have been counted. And there are mare areas with crater densities significantly lower than those at the youngest Apollo sites. The least densely cratered mare areas appear to be about 2.6 billion years old. Thus, during almost the first half of its life, the moon was hot inside and oozed basaltic lava. Since that time the moon has been geologically quiescent, at least to all outward appearances.

What about the older, far more densely cratered crust of the moon? Some kind of light-colored rock was already there when the basalts erupted; this rock now underlies the solidified basalt flows and pokes up between the basins to form mountain ranges standing as high as 5,000 meters above the mare surfaces. From the outset, scientists had great interest in the terra (highland) samples, because clearly the oldest samples of the moon would be among them.

As NASA mission planners gained confidence in the reliability of the Apollo system and the ability of the astronauts to maneuver into a safe landing area, they became more willing to send missions to the rough terrain of the lunar highlands. The last four Apollos landed in or adjacent to the highlands. As the program progressed, the capacity of an Apollo mission to carry out scientific exploration increased dramatically. It is interesting to make comparisons between the first (Apollo 11) and last (Apollo 17) missions:

- The time the astronauts spent on the lunar surface outside the Lunar Module (LM) increased from 2.2 to 22 hours.
- The distance from the LM they were able to explore and sample increased from 40 meters to 8 kilometers. (On the last three missions, the astronauts were supplied with an electrically powered "rover" to extend their range.)
- The weight of lunar samples they were able to collect and return to Earth increased from 21.7 to 110.5 kilograms per mission.
- On the last three missions, the Command and Service Module (CSM)—the spacecraft containing one astronaut that remained in lunar orbit while the LM and the other two astronauts were on the surface of the moon—carried an array of remote sensing instruments that scanned the areas it overflew.
- Most important, perhaps, the astronauts on the last three missions were far more highly trained and motivated to carry out scientific exploration than the earlier crews, who were necessarily preoccupied with testing the operational properties of the Apollo system.

During the later missions I was a member of the Lunar Sample Analysis Planning Team, an advisory committee that recommended which of the new lunar samples ought to be sent to the various approved lunar investigators for study. We also kept a suspicious eye on the handling and storage of the samples, attempting to modify or prohibit procedures that might contaminate or waste them or otherwise diminish their scientific value. We convened in Houston every month or two, meeting in a double house trailer in a platoon of similar structures. I was among the first people to see many of the samples unpacked from the containers loaded by the astronauts—an intoxicating experience.

The highland samples were extremely diverse in character, but they had a common property that was totally unexpected: a large proportion of the mineral plagioclase feldspar. Many of these rocks would have been called anorthosites or anorthositic gabbros if they had been encountered on Earth. (Anorthosite is just a rock containing a large percentage of plagioclase, a common mineral with a composition that lies somewhere between $NaAlSi_3O_8$ and $CaAl_2Si_2O_8$.) Why was this a surprise? What would one have expected the lunar surface to be composed of, in pre-Apollo times? There were several leading possibilities. The lunar surface might still be composed of unchanged, primitive planetary material (chondritic meteorites, as Urey speculated). It might be composed of basalts, more or less different than the mare basalts. (On the Earth's surface, basalt is the most widespread and ubiquitous of rock types.) It might conceivably have a composition similar to that of terrestrial andesites (volcanic rocks somewhat similar to basalts, but with systematic differences in composition and occurrence). Maybe granite was common on the moon. (Many of the major mountains on Earth are made of granite.) Or, perhaps the lunar surface material contained a component derived from its mantle (the subcrustal rock that was the source region for mare basalts). What these major possible rock types have in common is that they are either partial melts (basalt, andesite, and granite) or material that is undifferentiated except perhaps for the subtraction of a small amount of partial melt (chondritic and mantle rocks). Partial melting is understood to be the dominant igneous process on Earth. Anorthosite, which is a relatively uncommon rock type on Earth, is not created by partial melting, and this accounts for its absence from the list of likely lunar crustal materials.

On Earth, anorthosite is formed in comparatively small amounts by an igneous process called crystal fractionation. This process depends on the fact that, when a mineral crystallizes from a cooling melt, its density is not the same as that of the residual liquid. It may be denser than the liquid, in which case it tends to sink and accumulate at the base of the volume of magma. (Less likely, it may be lighter than the liquid, and tend to float.) Once the system is totally solidified, a sample broken or eroded from its base will be found to consist largely of that one mineral which was sinking and accumulating. If the mineral was plagioclase feldspar, then the rock where it accumulated is anorthosite.

This is the only process known to concentrate plagioclase. On Earth, however, it works only on a relatively small scale, and the anorthosite

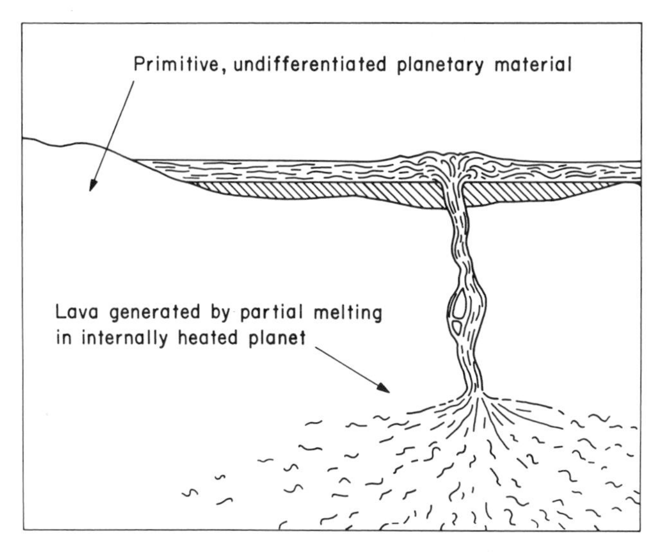

Figure 3.4
The two major possible sources of lunar rock, from a pre-Apollo perspective conditioned by terrestrial geology. Neither source would produce significant amounts of anorthosite.

layers are interspersed with cumulate layers having quite different compositions. Thus, it is understandable that this rock type was not a pre-Apollo nominee for the substance of the lunar crust.

Why, then, does anorthosite predominate on the lunar surface? If crystal fractionation is the only way of forming this rock, we have to picture an igneous event quite different than any known on Earth—one capable of accumulating plagioclase on a grand scale, and at the top of the melt system rather than at its base, since the anorthosite is observed at the surface of the moon. This must mean that plagioclase was lighter than the residual lunar melt system rather than denser, so it tended to float up and concentrate near the surface instead of sinking.

It is important to estimate how much lunar anorthosite we are talking about: All over the moon? And how thick a layer? The first question was answered by measurements made by the instruments in the orbiting CSMs of Apollos 15 and 16. An x-ray fluorescence detector and a gamma-ray detector measured the abundances of certain chemical elements in the lunar surface areas overflown by the CSMs. In all cases, whenever the CSMs passed over highland regions the chemical abundances detected were consistent with a more or less anorthositic composition. In addition, the astronauts on the lunar surface set up geophysical stations, which included seismometers (instruments that detect minuscule vibrations of the ground). Over a period of time, these instruments detected (and transmitted to Earth) the arrival of seismic waves created by moonquakes, meteoroid impacts, and spent lunar spacecraft that were purposely crashed into the lunar surface. From the travel times of seismic waves from each impact point to the various seismometers, the structure of the outermost layers of the moon could be deduced. A major discontinuity of physical properties was found at a depth of about 60 kilometers on the moon's near side. This is understood to be the base of the crust and the beginning of the mantle.

Is the whole thickness of the crust made of anorthositic rock? At first we thought it must be. The highlands had been so intensely cratered that materials of the upper and lower levels must have been thoroughly mixed, and the lower half of the crust could not be substantially different from what we observed in the upper half. However, later and more thoughtful study of the cratering history of the highlands revised this picture. There is no assurance that the highlands, in most areas, are cratered and mixed more deeply than a few kilometers. The lower crust could, it appears, be composed of something quite different from anorthosite.

But yet another item of evidence remains to be considered: What holds up the lunar highlands? Rock is more or less plastic over long periods of time; why didn't the highlands sink down to a common level with the maria? This may sound like a bizarre question, but mountains really do have to be supported by something. There are two possibilities. The lunar crust might actually be rigid enough to support the highlands by brute force. (If it were relatively cool, this would have the effect of increasing its strength.) Or the highlands might be "floating" in buoyant equilibrium in a denser material. This situation would be analogous to a log floating in a pond. The wood of the log is slightly less dense than the water, so it floats and projects slightly above the surface; the small amount of log above the surface is supported by the buoyancy of the much larger amount of wood submerged beneath it. In the lunar analogy, the several kilometers of anorthositic highlands that project as mountains above the mare surfaces would have to be buoyed up by a much greater thickness of relatively low-density rock.

Can a choice be made between these two possibilities? As it turns out, the answer is yes. The two models would have quite different effects on the gravity field near the lunar surface. If the highlands were held up by brute force, the extra mass contained in them would enhance the force of lunar gravity immediately above them. On the other hand, if the highlands "float," the gravitational effect of their extra mass would be offset by the deficiency of mass in the low-density layer that underlies them. In the first case, a spacecraft orbiting the moon would feel an extra gravitational tug when it passed over a highland area, and its orbit would be altered. In the second case this would not happen. A number of spacecraft orbited the moon before and during Apollo, and they did not experience the gravitational tugs that would signify nonbuoyantly supported highlands. The highlands are floating, and they must be underlain by a thick layer of relatively light rock. The only common low-density lunar mineral is plagioclase feldspar. Therefore, the 60 kilometers or so of near-side lunar crust must contain abundant plagioclase, and it is anorthositic in character.

How did this enormous quantity of one particular mineral come to be concentrated near the moon's surface? The only mechanism of concentration we know of on Earth, crystal fractionation, operates in pockets of melted rock (magma bodies) which are of limited size. The anorthosite layers that form in them are necessarily also of a scale that is negligible in global terms. We are forced to the conclusion that a magma body of enormous dimensions must have existed on the early moon to permit the separation of a thick anorthositic crust. This magma ocean, as it has come to be called, completely covered the moon and must have been several hundred kilometers deep. As it cooled and crystallized, the light plagioclase crystals tended to float upward and concentrate in a layer about 60 kilometers thick, which became the crust. The dense minerals that formed, olivine and pyroxene, sank and concentrated in the lunar upper mantle. Later this would become the source region of the mare basalts. Now we see why the mare basalts contain isotopic and trace element patterns that point to at least one cycle of previous igneous activity for their source regions.

The concept of a hot, melted surface region on the moon when it formed (or soon afterward) took the Earth-science community by surprise. The pre-Apollo consensus had been that the planets had probably formed in a relatively cool and undifferentiated condition, and that their interiors had gradually warmed up as a result of the radioactive decay of uranium, thorium, and potassium 40. Now, suddenly, a radically different beginning was demonstrated for the moon. It was understood that the energy to heat and melt the outer layers of the moon was probably associated with its accretion. That is, the planetesimals that joined the growing moon smashed into it with high

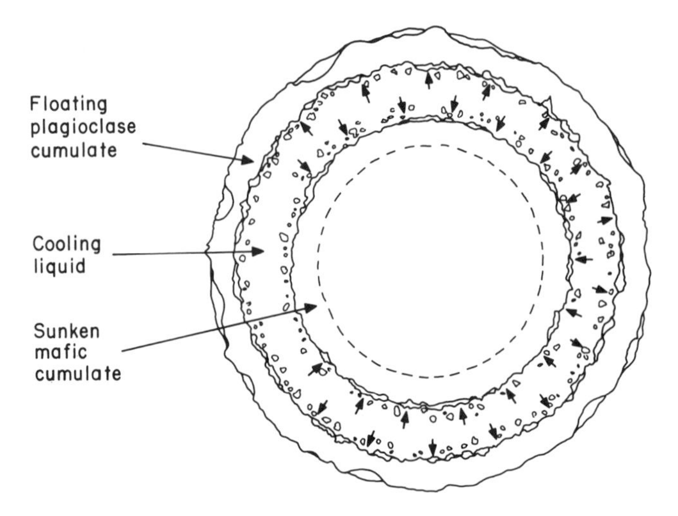

Figure 3.5
The hypothetical lunar magma ocean, shown in a highly schematic diagram. Inside the dashed line is unmelted primitive lunar material; outside it the magma ocean is cooling and crystallizing. Dense olivine and pyroxene sink and join the upper mantle; light plagioclase floats and joins the crust.

velocity, and most of their kinetic energy was converted into heat energy. There were and still are problems in accounting for the early thermal state of the moon, but the evidence of extensive early melting is clear. The concept of a hot origin has since spread to all the inner planets. Current discussions of the earliest evolution of Earth are usually in the context of a magma ocean (figure 3.5).

This has been very much a broad-brush treatment of the lunar highlands. Only the most general properties of the rocks (their high content of plagioclase feldspar) and the outline of an interpretation (the magma ocean) have been alluded to. Based on our experience with terrestrial rocks, we might expect to learn a great deal more about the situation from detailed studies of the chemical, isotopic, and petrographic properties of individual samples. However, it has turned out to be rather difficult to do so. This was foreseeable. Viewed from a distance, the lunar highlands are a tortured landscape of craters overlapping craters, not unlike pictures one sees of World War I battlefields. The highland rock must have been horribly smashed up and indiscriminately mixed by all those cratering impacts. Could any of the original properties of the highland rock have been preserved in a readable form?

As expected, many of the highland rock samples turned out to be welded-together aggregations of rock and mineral fragments, the debris

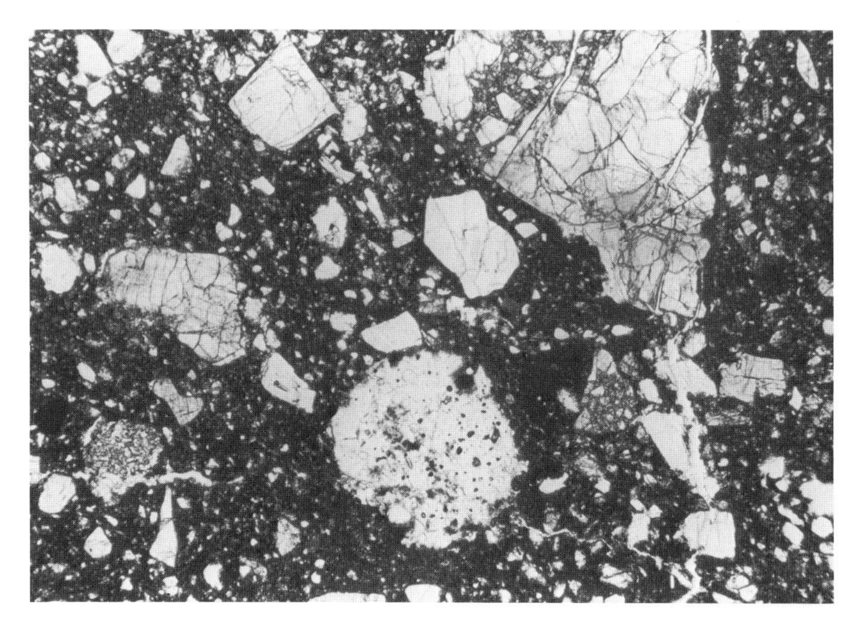

Figure 3.6
Microscope view of a thin section of a breccia sample (rock 14321) collected on the Apollo 14 mission. The field of view is 3 millimeters wide. The lightest areas are the most transparent. The rock consists of angular fragments of earlier generations of rock (composed mostly of plagioclase) embedded in darker, fine-grained, partly glassy matrix. (Smithsonian Astrophysical Observatory photograph)

of countless cratering impacts. (These aggregations are called breccias.) Clearly the bulk properties of breccia samples are of very little scientific value, since they average together the properties of many earlier generations of rock. However, there were also igneous rocks, and many of the fragments in the breccias that were large enough to be extracted and studied were of igneous rocks. These fragments have been studied intensively. Some are variants of the basalt clan, and these testify to the nature of pre-mare lunar volcanism. Many of the other lunar igneous rocks turned out to be not all that informative. Some of us were a bit naive in thinking that igneous rocks containing microscopic crystals might have been products of the great magma ocean; actually, crystals that formed in a body that vast would more likely have grown to the size of Volkswagens. What we did not appreciate at the time was that high-energy cratering events not only bash up rock and toss it around, they can also melt it.

Laboratory experiments in the 1960s that simulated lunar cratering had indicated that relatively little target material was melted. As our understanding of cratering mechanics improved in the 1970s, however, it became clear that large cratering events cause extensive melting. The larger the impact, the more melting occurs—not only in absolute quantities, but the proportion of target material that is melted instead of being broken up also increases. Much of this understanding came from field studies of huge ancient impact structures on Earth, especially the 66-kilometer Manicouagan crater in northern Quebec, where great thicknesses of solidified melt rock were found.

Thus, a large proportion of the lunar igneous rocks are now understood to be impact melts, and the target materials that were melted were undoubtedly as mixed up and generally uninformative as the bulk breccia samples. Much of the continuing research on lunar samples in recent years has focused on identifying the few highland samples that appear to be surviving "pristine" crustal material and trying to infer details about the magma ocean from them.

What about the ages of highland samples, pristine and otherwise? These rocks must have formed and been in place before the mare basalts began to erupt, 3.9 billion years ago. Radiometric dating of the highland rocks yielded a surprising result: Their potassium-argon ages are clustered strongly in the age range 3.9–4.0 billion years; that is, back to just before the epoch of mare basalt eruption. It is very unlikely that these rocks actually took up their present chemical compositions at that time. In fact, there is in them subtle evidence from other isotopic systems of earlier chemical fractionations, probably associated with the crystallization of the magma ocean. Remember, though, that the potassium-argon method depends on the retention of radiogenic argon atoms inside mineral crystals, and argon cannot accumulate in a rock that is hot. Thus, if a rock (no matter how old) is heated beyond a certain temperature, its accumulated argon is driven off and "its potassium-argon clock is reset to zero," as the saying goes. The clustering of potassium-argon ages at 3.9–4.0 billion years has been widely interpreted to mean that some high-temperature event or condition at that time reset the clocks of highland rocks all over the moon, regardless of how old they really were, to zero.

Many of the highland samples are undoubtedly pieces of crustal debris that were thrown out of the Imbrium basin (the gigantic circular crater, now filled with lava, that occupies much of the upper left quarter of the full moon) when a large impacting planetesimal blasted it out. Indeed, the Apollo 14 and 15 sites were chosen expressly to sample what appeared to be Imbrium ejecta. Much of the debris from such a monstrous explosion undoubtedly would have been made hot

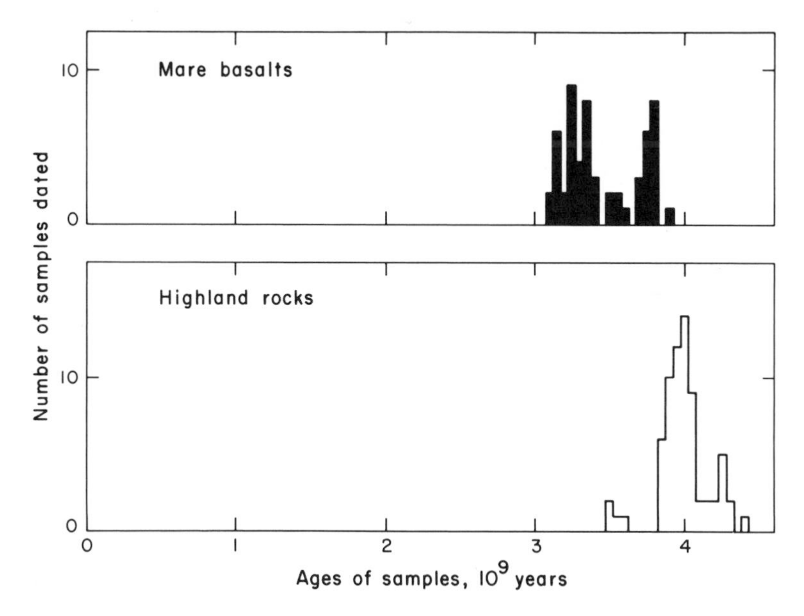

Figure 3.7
Potassium-argon ages of mare and highland rocks. (Compilation of data courtesy of G. Turner)

enough to outgas argon. Thus, we would expect to find a concentration of rock ages recording the time when the Imbrium impact occurred.

Surprisingly, however, samples from the Apollo 16 and 17 missions, which landed on the ejecta blankets of the Nectaris and Serenitatis basins respectively, showed essentially the same concentrations of ages at 3.9–4.0 billion years. Therefore, a concept emerged that virtually all the visible circular mare basins were blasted out of the moon in a very short time by a concentrated bombardment of major planetesimals, and this time of torment was left imprinted in the ages of the great majority of highland rocks. "It must have been a great show," commented lunar scientist Jerry Wasserburg, "if you had a pretty good bunker to watch it from." Needless to say, the nearby Earth would have shared any such bombardment that the moon experienced.

Actually, there are two variants of this interpretation. One is that the "terminal cataclysm" was a concentrated bombardment that followed a period of relatively quiescent crust formation, resetting the ages of most of the crustal rocks. The other is that bombardment was even more severe before the "terminal cataclysm," that innumerable huge basins were excavated (which we can't even see any more), and that crustal rocks never had a chance to accumulate any argon until after 4.0 billion years ago. In either case, the bombardment rate by planetesimals large and small declined greatly after 3.9 billion years ago (presumably because the huge number of planetesimals that had been swirling about in the early solar system were finally swept up by the planets), and just at that time floods of basaltic lava began to well up through cracks in the lunar crust and flow into the newly excavated impact basins.

Much of the picture I have sketched was developed in the early 1970s, during and soon after the Apollo program. Many major questions remain open, but the answers come more slowly now. Possibly the most important recent discovery has been the observation that there are certain systematic differences in chemical composition among the pristine samples according to the longitude of their occurrence on the moon. Western samples (Apollos 12, 14) differ from central samples (Apollos 11, 15, 16, 17), and these differ from eastern samples collected by three Soviet unmanned spacecraft (Luna 16, 20, 24). This opens the question of lateral heterogeneities in the composition of the moon. Actually it is not the first indication of asymmetries in the moon. An earlier one was the discovery by the gamma-ray detectors on orbiting CSMs that radioactive elements such as uranium are strongly concentrated in the vicinity of Mare Imbrium and relatively rare elsewhere. Another was the observation that there are virtually no maria on the far side of the moon. (This was discovered in 1959 by a crude Soviet spacecraft, Luna 3, which flew behind the moon and transmitted back blurry pictures of the far side.) At first we wondered why basin-forming planetesimal impacts had occurred on only one side of the moon. In time, however, lunar photogeologists pointed out that there are as many major impact basins on the far side as on the near side, but no basalt flowed into the far side's basins.

Another type of asymmetry was found by the altimeter that continually measured the height of the orbiting CSM above the surface by beaming laser pulses downward and measuring the time it took for them to bounce back. Of course, this experiment found a very irregular surface, but if we smooth out the irregularities we are left with an interesting major discrepancy: The smoothed surface of the near side stands closer to the center of mass of the moon, by several kilometers, than the smoothed far side. The significance of this is most simply stated as follows. If we could slice the moon into two hemispheres (near side and far side) through its center of mass, the far-side hemisphere would be slightly larger in volume than the near-side hemisphere, since its mean surface stands farther from the center of mass than the near side's surface. Therefore, the mean density of the substance of the far side is slightly less than that of the near side.

The original interpretation of this fact was that the plagioclase-rich crust of the moon must be systematically thicker on the far side (roughly 100 versus 60 kilometers). This would rationalize the elevation difference between the near and far sides. A thicker "log" of anorthosite, floating in the "pond" of denser lunar mantle material, would project a little higher above the surface of the pond. This model also helps explain the absence of basalt flows on the far side: Maybe basalt was generated in the mantle all around the moon, but, although the hydrostatic pressure exerted on the liquid was adequate to pump it up to the surface on the near side, this was not enough pressure to raise it a few extra kilometers on the far side. (The reason for a systematic asymmetry of crustal thickness on the moon was never given.) But models of this sort start with the assumption of uniform material compositions and densities all around the moon. Once evidence of lateral heterogeneity appears, the picture becomes much more complex. The lower mean density of the far side could now be understood as due to lower-density material in the far-side mantle or crust or both, rather than to a greater amount of low-density crust. A gratifyingly simple explanation is lost, but we gain glimpses of a tantalizing question: Why would there be lateral variations in chemistry?

I have evaded the question of the origin of the moon. The Apollo program did not contribute greatly to the solution of this problem, which remains open. One possibility is that planetesimals in the early solar system, swirling past the growing Earth, tended to be captured by some deceleration mechanism into a swarm of bodies orbiting the Earth. In time, the swarm coalesced into the moon. There are many problems and difficulties with this model. The other major possibility is that the moon came out of the Earth. It has been argued that similarities in the abundances of certain trace elements in terrestrial and lunar basalts must mean that their source regions (the terrestrial and lunar mantles) had a common origin. The model proffered is that a very large planetesimal hit the growing Earth a glancing blow, knocking a quantity of mantle and early crustal material away from it. However, orbital mechanics dictates that such debris would either escape the Earth altogether and go elsewhere in the solar system, or, after making one circuit about the Earth, it would fall back onto its surface. Problems here, too.

Apollo greatly advanced our understanding of the other bodies of the inner solar system. It demolished the notion that other planets would be dramatically different from Earth, and it underlined the fact that the same laws of physics and thermodynamics produce essentially the same minerals, rocks, and geological processes everywhere. It confirmed the universal importance of a particular rock type, basalt, and at the same time gave us a new perspective on the initial state of planets (hot) and the first separation of crustal material.

Yet the moon remains an enigma, a baleful disk that hangs soft and orange as cheese over the horizon or brittle and bright blueish-white high in the midnight sky. It may be many years before another major advance in lunar science. Efforts to promote even the most modest post-Apollo scientific mission to the moon have failed. Perhaps we will have to wait for another time when there is a compelling nonscientific reason to return to the moon, to carry science back there again on the coattails of some great venture.

4

Voyager to the Giant Planets

Bradford A. Smith

In late August 1981, a semi-intelligent robot sped swiftly away from the giant ringed planet Saturn, moving ever deeper into the cold, dimly lit reaches of the outer solar system. Five years and 2 billion kilometers earlier, the small spacecraft had been lifted from Florida's Kennedy Space Center by a Titan-Centaur launch vehicle to begin its historic voyage to Jupiter, Saturn, and beyond. So rapidly was it hurled from Earth that it crossed the orbit of the moon only 8 hours after the booster's final push. The spacecraft was Voyager 2, and along with its sister ship Voyager 1 it would provide mankind's first close look at that remote and poorly understood region of the solar system that lies beyond the asteroid belt. Now, leaving Saturn behind, Voyager was showing signs of its advancing age. Arthritic, somewhat deaf, and with growing indications of senility, the machine was beginning its long trip through the void of interplanetary space to Uranus and Neptune. We do not know if Voyager 2 can survive those additional 41/2–8 years in space, but whether or not it succeeds Voyager has already accomplished most of its designers' goals. We have seen some of the bizarre worlds of the outer solar system, and we are learning how they fit into the overall picture of the solar system's creation and evolution.

The Voyager project is but one phase in a program of planetary exploration—a program designed to help us better understand how our solar system came into existence, how it has evolved from those early years, and ultimately how life was created. Indeed, the answers to questions relating to the origin of life may eventually be found not only in biochemical laboratories here on Earth, but elsewhere in the solar system. Unfortunately, as a result of nearsighted fiscal policies of the recent and current administrations, this program of planetary exploration is today in great jeopardy. Twenty years of intensive study of our planetary neighbors may be drawing to a close. The nation that led the world in space exploration seems now ready to relinquish that role to the Soviet Union, Western Europe, and Japan. But this is not the forum to dwell on the current problems of the American space program; instead we should reflect on our accomplishments as we close out these two extraordinary decades of solar-system exploration.

We began this program by exploring the planetary bodies of the inner solar system: Mercury, Venus, the moon, and Mars and its satellites. But less than 10 years after our first spacecraft were sent to the moon, we began directing our interests beyond the orbit of Mars toward the giant planets: Jupiter, Saturn, Uranus, and Neptune.

Jupiter, 11 times the diameter of Earth and over 1,300 times Earth's volume, contains more mass than all the other planets combined. Saturn is more than 9 times the size of Earth, and with its magnificent set of rings its overall extent spans two-thirds the dimension of the entire Earth-moon system. Uranus and Neptune are each about 4 times larger than Earth. Before the space age, our knowledge of these planets—especially Uranus and Neptune—was abysmally poor. This is understandable; they are very far away. Jupiter's average distance from Earth is nearly 780 million kilometers, Saturn's 1.5 billion kilometers, Uranus's 2.9 billion kilometers, and Neptune's 4.5 billion kilometers. Much of what we did know involved composition, for compositional measurements are not especially sensitive to the remoteness of the objects being studied. All the giant planets are composed primarily of hydrogen, with some lesser amount (approximately 10 percent) of helium. Still smaller quantities of carbon, nitrogen, oxygen, and sulfur compounds are found in their atmospheres, mostly in the form of methane, ammonia, ammonium hydrosulfide, water, and various organic compounds.

As we observe the surfaces of the giant planets, we see only layers of clouds floating in cold, hydrogen-rich atmospheres; below, the atmospheres become compressed to liquid hydrogen, and at the centers of these planets there may be small cores of silicates and metals. With the unexplained exception of Uranus, all the giant planets have internal heat sources—that is, energy left over from their creation more than 4.5 billion years ago. Because their atmospheres receive thermal energy, which comes not only from the sun but from their interiors as well, these planets have meteorologies very different from those of the inner planets. Jupiter and Saturn show bands of colored clouds drawn out parallel to their equators. Such patterns imply the presence of strong

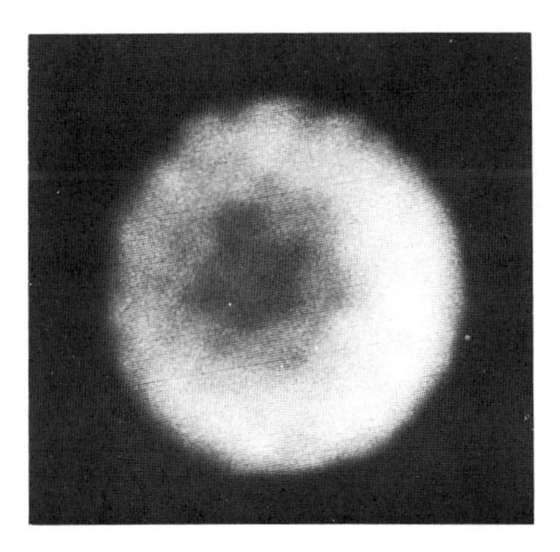

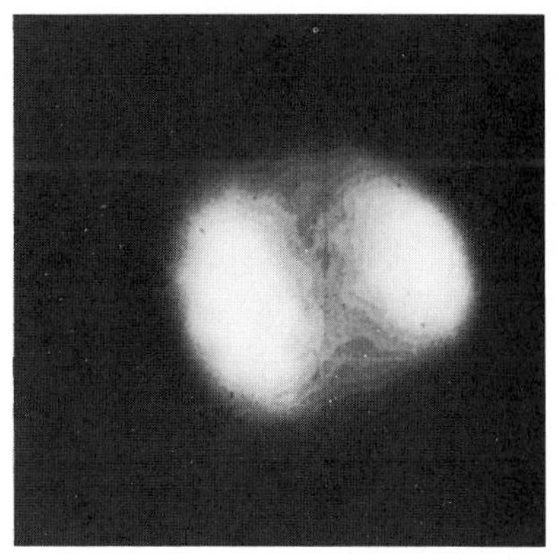

Figure 4.1
(Right) View of Uranus showing high clouds that brighten toward the limb. (Left) View of Neptune, also showing high clouds. Both planets were imaged at the 0.9-micron absorption band of methane using a CCD detector. (Courtesy of Lunar and Planetary Laboratory, University of Arizona)

zonal (east-west) winds, which has been confirmed by ground-based tracking of discrete clouds in Jupiter's atmosphere. Jupiter's atmospheric cloud structure is more variegated than Saturn's, with colors that range through brick red, orange, white, brown, and mauve. Saturn's clouds, which are yellow to orange and more uniform, are rarely seen as the discrete features so necessary to the mapping of global atmospheric patterns. Because of their great distances from Earth, we know very little about the cloud structures on Uranus and Neptune; each exhibits a pale bluish-green disk caused by the strong absorption of red light by relatively large amounts of atmospheric methane.

Of the nearly 50 satellites known in the solar system, all but three (Earth's moon and the two tiny Martian satellites) are in the outer solar system. Jupiter is known to have 16, Saturn at least 21, Uranus 5, and Neptune at least 2. Most of these satellites are small (less than 1,500 kilometers in diameter), but the four Galilean satellites of Jupiter and Saturn's largest satellite, Titan, are planet-size bodies. With the possible exception of Neptune, all the giant planets are known to have rings. Saturn's rings have been known since the earliest telescopic observations, the Uranian rings were discovered in 1977 during an occultation of a star by that planet, and Jupiter's rings were found by Voyager in 1979.

The exploration of the outer solar system began when Pioneer 10 reached Jupiter in late 1973. Pioneer 11 reached Jupiter just a year later and went on to encounter Saturn in 1979. Pioneer was a relatively simple, spin-stabilized spacecraft designed to act as a pathfinder for the more complex Voyagers that were to follow. Nevertheless, important discoveries were made by Pioneer. One of the most important was that of an intense radiation belt surrounding Jupiter, similar to the Van Allen belts around Earth but many thousands of times stronger. Luckily, the Pioneers survived their passages through the Jovian environment; without benefit of the warning from the Pioneers and a subsequent redesign, the Voyagers would have been cooked by the overwhelming radiation. The Pioneers also gave us a better view of Jupiter's complex cloud structure, including the Great Red Spot, and discovered a narrow ring surrounding the bright ring system of Saturn. But because the optics in Pioneer's imaging system had a relatively short focal length, resolution superior to that attainable with the better ground-based telescopes was not achieved until just a few days before the Jupiter and Saturn encounters. Thus, there was not enough time to record data for dynamical studies of Jupiter's atmosphere, and the combined data from ground-based telescopes and from Pioneer could only hint that the meteorology of Jupiter might be every bit as complex as its cloud structure appeared to be. Pioneer failed to record any discrete cloud features on Saturn, thereby creating grave concern within the Voyager Imaging Team over our prospects for studying the dynamical regimes of Saturn's global circulation. It had been our hope that the detection of discrete clouds on Saturn lay just below the resolution attainable with ground-based telescopes, and that Pioneer and Voyager would reveal a wealth of cloud structure. Our initial disappointment over the Pioneer images was later to be reinforced by the early sequence of "engineering" and calibration images of Saturn taken by Voyager even as late as 6 months before the first Saturn encounter.

The Voyager project, initially referred to as "Mariner Jupiter-Saturn," was created by NASA in 1972 as a "follow-on" to the Pioneer program. The Voyager spacecraft is, in fact, of the Mariner class: a relatively large, three-axis-stabilized spacecraft, capable of operating in deep space on stored commands issued periodically from Caltech's Jet Propulsion Laboratory in Pasadena, California. Both Voyager spacecraft were launched in the late summer of 1977 from NASA's Kennedy Space Center. Voyager 2, which was to take a slower trajectory to Jupiter and Saturn, was actually launched first, on August 20; Voyager 1 followed on September 5 and soon overtook its sister spacecraft. As the two Voyagers roared upward from the Florida coast atop their Titan-Centaur launch vehicles, they began a journey that would eventually take them beyond the solar system. Voyager 1 would visit both Jupiter and Saturn, then head out toward the stars. Voyager 2 had within its trajectory a built-in option to go on to Uranus and Neptune before leaving the solar system; the decision whether to exercise this option would not be made until Voyager 1 had successfully completed the first Saturn encounter. The Titan-Centaurs, powerful and reliable, were ideal machines for dispatching NASA's exploratory robots to the far corners of the solar system. Unfortunately, the Voyager Titan-Centaurs were the last of their kind in NASA's arsenal of launch vehicles, and all future missions to the outer solar system are, for better or for worse, tied to the NASA Space Transportation System, known more popularly as the Space Shuttle.

Shortly after New Year's Day 1979, the Voyager Imaging Team began preparing for the first Jupiter encounter, which was to occur on March 5. Already the first of the spacecraft's programmed imaging of Jupiter was appearing on our television monitors. Images in several spectral colors were being received every 2 hours, around the clock, 7 days per week. As had been suggested by the Pioneer images, the individual cloud systems in Jupiter's atmosphere exhibited a very complex structure; however, the complexity exceeded our expectations. With each passing day the images grew larger and, with improving resolution, the cloud structure appeared more complex. We began to suspect that such a complex cloud structure might indicate equally complicated dynamical characteristics, and we would soon learn that our concern was warranted.

Jupiter, the giant of the solar system, loomed before us on our monitors showing cloud color, structure, and dynamical regimes that we could never have imagined. The simple zonal motion inferred from the ground-based and Pioneer observations was instead seen to be the net result of complicated cyclonic and anticyclonic vortices, similar to high- and low-pressure systems in the Earth's atmosphere. Some cloud systems seemed to consume and then regurgitate their neighbors. Others circled one another before moving apart and drifting away on separate currents. Convective clouds appeared from below, forming great white patches as they came into view, only to be sheered out into long cometlike features by adjacent wind currents. The Great Red Spot, known to astronomers since the middle of the seventeenth century, was found to be a giant vortex, an anticyclonic high-pressure system large enough to enclose three Earths.

Early in the approach of each Voyager, a number of individual cloud systems in Jupiter's atmosphere, including the Great Red Spot, were selected for examination at high resolution near the time of

Figure 4.2
Jupiter as seen by the approaching Voyager 1. Convective features stretched by wind sheer can be seen in the equatorial region. (NASA photograph)

Figure 4.3
Detail of the Great Red Spot of Jupiter recorded by Voyager 1. (NASA photograph)

closest approach. This required us to make a long-range Jovian weather forecast; we had to determine whether a particular system would still exist more than a month later and, if so, exactly where it would be. Both ground-based and spacecraft observations were employed in making these forecasts, which turned out to be 95 percent accurate.

Observations of Jupiter's nighttime hemisphere revealed auroras near the planet's north pole and more than a dozen lightning bolts, some of which were hundreds of times more intense than those that occur on Earth. Lightning on Jupiter was not unexpected, for we had long believed that large convective systems, similar to terrestrial thunderstorms, abounded in the planet's atmosphere. What was unexpected was Jupiter's ring. Theory had argued against Jupiter's ability to retain a planetary ring, but one does not travel nearly a billion kilometers into space without at least taking a look. We looked. It was there, one of the most startling discoveries of the Voyager mission. Jupiter's ring bears little resemblance to the well-known rings of Saturn. It is very narrow and exceedingly faint, barely detectable (we now know) with ground-based telescopes. The space between the ring and Jupiter is filled with tiny particles, spiraling inward and eventually deposited in the upper layers of the planet's atmosphere. If ring particles are thus

Figure 4.4
Voyager images of the Galilean satellites of Jupiter, scaled to size. Clockwise from upper left: Io, Europa, Callisto, Ganymede. (NASA photographs)

being continuously depleted, there must be a source. Interplanetary dust? The grinding up of two small satellites found imbedded within the ring? Perhaps. We still know very little about the mechanisms that operate in Jupiter's ring.

Of special interest to the Voyager Imaging Team were the four Galilean satellites (figure 4.4). All are larger than Pluto, and Ganymede and Callisto are comparable in size to the planet Mercury. Thus, we tend to think of these bodies as planets, although their association with a major planet defines them as satellites. Both Ganymede and Callisto were found to be composed of water-ice shells surrounding a core of silicate. Callisto showed an ancient dusky surface, darkened by eons of infalling interplanetary dust but with fresh, bright patches created by relatively recent meteorite impacts. The surface of Callisto appears about as old as any in the solar system, and since nearly all the bodies in the solar system were created at about the same time

(4.6 billion years ago) this means that Callisto's surface has changed very little since the satellite's formation. In other words, Callisto has been geologically dead since birth. Ganymede, the largest satellite in the solar system, bears a strong resemblance to Callisto, although several regions have undergone tectonic modifications (faulting and folding) in a process similar to plate tectonics on Earth. The major difference, of course, is that the terrestrial tectonics involve motions and interactions of huge plates of solid silicate, whereas on Ganymede the material is water ice with its upper layers made rigid as steel by the extreme cold.

Europa and Io are the smallest of the Galilean satellites, roughly comparable in size to Earth's moon. Europa is mostly silicate surrounded by a thin shell of water ice. The surface of Europa is remarkably smooth, showing very few impact craters, but covered by a network of narrow, dark ridges scarcely more than 100 meters high. The absence of any significant topography suggests that the surface is relatively young. Perhaps internal heat sources have kept the surface warm enough to allow the ice to flow like terrestrial glaciers, thereby destroying impact features as rapidly as they are created. The network of linear and sinusoidal ridges remains a mystery.

The one Galilean satellite not covered by ice is Io, which we had expected to look like a reddish version of our own moon, pockmarked with thousands of meteorite craters. Although no images of Io taken by ground-based telescopes or by Pioneer were able to show surface features, the spectrum of its reddish surface suggested that it was covered with sulfur. In contrast with the three other Galilean satellites, its spectrum did not show even a trace of water. Prepared to see a "sulfur-dusted moon," we were totally unprepared for what we did see. Io showed not a single impact crater down to our resolution limit of one kilometer; its surface appeared to be not only very young, but the youngest ever seen in the solar system, including that of Earth. Our estimates placed its probably age at a few million years—the blink of an eye in comparison with the 4,600 million years that the solar system has existed. Great regions stretched out before us, but without the familiarity of recognizable geological features. Strange colors added to the disorientation. The entire surface appeared to have been ravaged by some horrible disease. Then we found the volcanos.

Ever since the earliest days of planetary exploration we had been searching for evidence of extraterrestrial volcanism. Volcanos had been seen on the moon, Mars, and Mercury, and probably on Venus, yet none of those appeared to be active. But Io had active volcanos;

Figure 4.5
Eruption of Maui (on the limb) and Prometheus volcanos on Io as observed by Voyager 1. Note the umbrella-shaped profile of Maui and the streamers of ejecta issuing from Prometheus. (NASA photograph)

at least a dozen of them were seen during the two Voyager encounters. Had Voyager flown past Earth at the same distance, it would have been fortunate to have recorded a single volcanic eruption. On Io, we had found not only active volcanism, but the most volcanically active planet known. Yet the volcanism on Io is very different from the terrestrial kind. Volcanism on Earth is produced when molten rock, or magna, heated by radioactive elements in the crust, is extruded from subsurface reservoirs. If the magna comes in contact with ground water, the rapid expansion of the water as it turns to steam produces an explosive form of volcanism. Similar mechanisms occur on Io, but the materials are different, the temperatures are lower, and the energy appears to come from strong crustal tides raised by Jupiter rather than from radiogenic heating. The magna on Io appears to be molten sulfur instead of silicate; temperatures are limited to approximately

400°K rather than the 1,200°K encountered in terrestrial silicate magmas. There is no water left on Io; all of it was presumably lost to space over eons of volcanic activity. Yet explosive volcanism is common on this Jovian moon. Liquid sulfur dioxide runs beneath the surface, and when it comes into contact with molten sulfur the results are spectacular (figure 4.5). Volcanic plumes have been seen rising more than 300 kilometers above the surface of Io, raining particles of solid sulfur and crystals of sulfur dioxide snow over the countryside. Such deposits quickly bury any impact topography. The surface of Io is as young as yesterday.

As the two Voyager spacecraft receded from Jupiter, they left behind a planetary system that would never again be viewed as before by either scientists or nonscientists. Although we did not understand all that we saw, the giant planet became curiously familiar to us, and the satellites, once merely points of light in the astronomer's telescope, had become real and individually recognizable worlds.

From Jupiter the two spacecraft sped toward Saturn, where they would arrive in mid-November 1980 and late August 1981. If we were surprised by what we saw at Jupiter, there would surely be further surprises at more-remote and less-understood Saturn.

The concern over the apparent absence of discrete clouds in Saturn's atmosphere vanished when systematic imaging began three months before encounter. The clouds were there, but they were generally smaller, and lower in contrast, than their Jovian counterparts. An approach sequence similar to that employed at Jupiter revealed wind patterns similar to those seen on Jupiter, but with different speeds and latitudinal extent. Jupiter has an equatorial jet stream, approximately 15,000 kilometers wide, which moves eastward at about 100 meters per second. Saturn's equatorial jet is 80,000 kilometers wide and has an eastward speed of nearly 500 meters per second. The reasons for these differences are not fully understood, but they may be due to the great differences in depth of the atmospheric circulation between the two planets. Several other jet streams were observed, some having longitudinal waves with high- and low-pressure systems nested in the peaks and troughs (figure 4.6). Dynamically and structurally, the Saturnian jets with their associated cyclones and anticyclones bear a startling resemblance to the terrestrial jet streams. The intensive study of the meteorology of another planet can be a great help in understanding and predicting the vicissitudes of our own.

The satellites of Saturn are generally smaller and more numerous than those of Jupiter. Titan (Saturn's largest satellite, only slightly smaller than Ganymede) must be considered one of the most intriguing

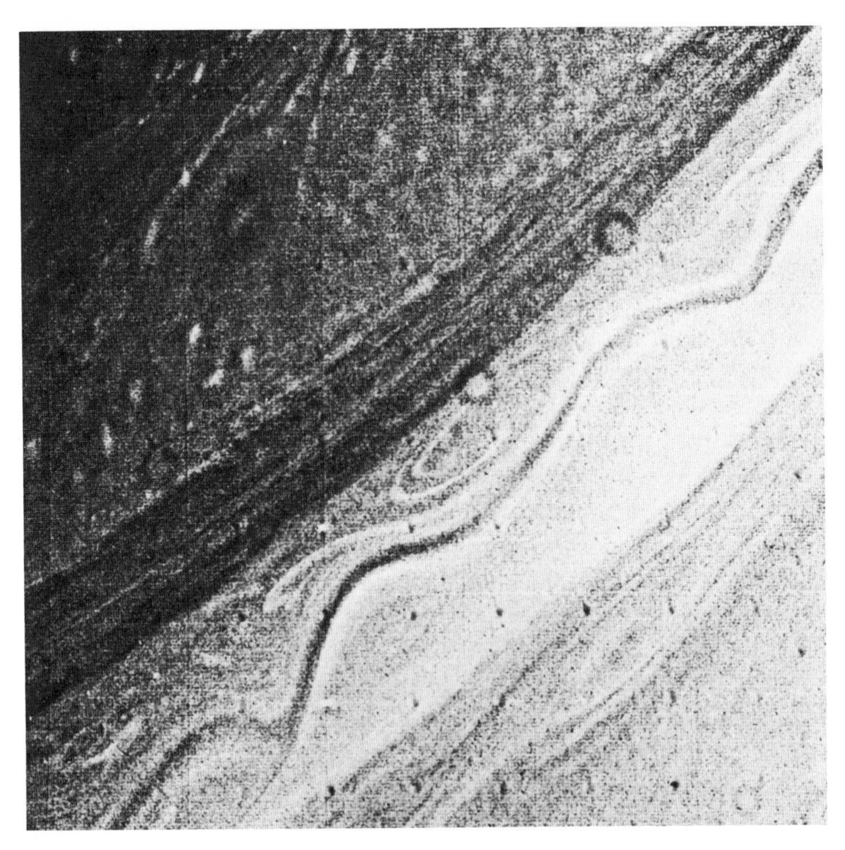

Figure 4.6
Waves in a Saturnian jet stream at mid-northern latitudes as observed by Voyager 2. Cyclones and anticyclones, analogous to low- and high-pressure systems, can be seen nested in the peaks and troughs of the waves. (NASA photograph)

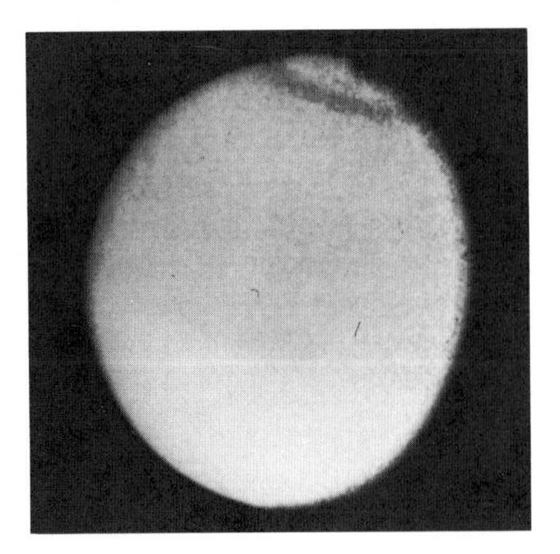

Figure 4.7
Titan as viewed by Voyager 1. Surface details are hidden by dense clouds. (NASA photograph)

Figure 4.8
Voyager images of the larger satellites of Saturn, scaled to size: Clockwise from upper left: Phoebe, Hyperion, Rhea, Tethys, Mimas, Enceladus, Dione, Iapetus. (NASA photographs)

satellites in the solar system. Titan is the only satellite known to possess an atmosphere, and its surface pressure is more than half again as great as Earth's. The composition of this atmosphere makes Titan even more interesting: nitrogen, methane, hydrogen cyanide, and other organic materials. This composition is very similar to that which existed on Earth nearly 4 billion years ago, when complex chemical evolution gave rise to life on our planet. Unfortunately, here on Earth, life changed forever the environment that had created it, and we are left with laboratory simulations to tell us what processes were taking place on Earth during those early years. The temperatures on Titan are around −200°C, too cold to create or sustain life. Nevertheless, many nonbiological chemical reactions similar to those of the early Earth must be going on even now in Titan's atmosphere, much as they have been for billions of years. The surface of Titan may well be covered with tens of meters of frozen organic sludge, which could answer many of our questions on just how life got started on Earth. Certainly one of the major objectives of future space exploration will be the landing of a Viking-type spacecraft on Titan's surface to perform an *in situ* chemical analysis.

The other satellites of Saturn, at least twenty in number, are small (10–1,500 kilometers in diameter). With the exception of Phoebe, they are all composed of water ice with lesser amounts of silicate. Phoebe, the anomalous outer satellite, is very dark, revolves about Saturn in a retrograde orbit, rotates in a prograde direction in about 10 hours, and is almost certainly a captured asteroid or degenerate comet nucleus. All the other Saturnian satellites, so far as is known, rotate synchronously; that is, their rotational and orbital periods are identical. With the exception of Titan, whose cloud-covered surface cannot be seen, all show the battered scars of impacts by asteroids, comets, and loose debris orbiting within the Saturn system.

Two of the smaller Saturnian satellites have generated considerable interest. Iapetus, the outermost regular satellite, is as dark as coal on one side and as bright as slightly dirty snow on the other. The dark hemisphere is precisely coincident with the satellite's leading orbital hemisphere. Both external and internal mechanisms have been proposed, but none has yet been satisfactory in explaining the observed distribution of the bright and the dark material. Enceladus, one of Saturn's closer satellites, has regions that are remarkably smooth and nearly featureless in a way that is quite reminiscent of Europa. Enceladus might be the site of a truly bizarre phenomenon: ice-and-water volcanism. The energy could be tidal in origin, much like the heat that drives Io's volcanos. Further support for this idea lies in the coincidence of a tenuous ring, called the E ring, with the orbit of Enceladus. The E ring appears to be dynamically unstable, losing material in the same manner as the Jovian ring. The source replenishing the lost ring particles could well be eruptions of these hypothetical water-ice volcanos on Enceladus.

During the planning of our encounters with Saturn, we prudently anticipated the unexpected. Saturn itself was an enigma, poorly seen through even the best of astronomical telescopes; the satellites were even less understood and had never been seen as anything more than points of light. It was with Saturn's rings that we felt the most comfortable. We thought we understood their structure and dynamical properties; surely there would be few surprises. We could not have been more wrong. Long before the closest approach by Voyager 1, the three main rings began to show internal structure, breaking down into hundreds and then thousands of individual ringlets. Voyager 2, making a closer approach to the ring system, showed that the total number of ringlets must exceed 10,000. Moreover, some of the ringlets were eccentric and variable in width, and others were discontinuous.

One of the most perplexing of the ring phenomena seen by Voyager was the "spokes," dark radial features extending across the B (middle) ring and revolving about Saturn with the ring particles (figure 4.10, top). When seen in forward-scattered light (that is, with the spacecraft's cameras looking in the general direction of the sun), the spokes became bright (figure 4.10, bottom). This photometric behavior suggested that the spokes were composed of very small particles, but this told us nothing about what caused them to appear and disappear. Although we still do not have all the answers, it seems very likely that the spokes are somehow associated with Saturn's electric field or magnetic field or both, and that the appearance and the lifetime of these features involve the interaction of very small, charged ring particles with the fields.

Another puzzling feature was the narrow F ring, discovered by Pioneer and located just outside the outermost ring (A) of the bright rings of Saturn. In the highest-resolution images (such as figure 4.11), this ring showed several components or strands that appeared to be kinked, knotted, and even twisted. The dynamical properties of this ring appear to be controlled, at least in part, by two small, nearby "shepherding" satellites that orbit just inside and just outside the F ring. Recent attempts to explain the structure of the F ring in terms of complex gravitational perturbations of the ring particles by the shepherding satellites have met with only qualified success.

Figure 4.9
Voyager photographs of Saturn's bright ring system. The upper photo shows the multiple ringlets at low resolution; the lower one shows a part of the B ring at high resolution. (NASA photographs)

Figure 4.10
Voyager 2 revealed the enigmatic "spokes" in Saturn's rings. Seen in back-scattered light, they are dark (upper photo); in forward-scattered light, they are bright (lower photo). (NASA photographs)

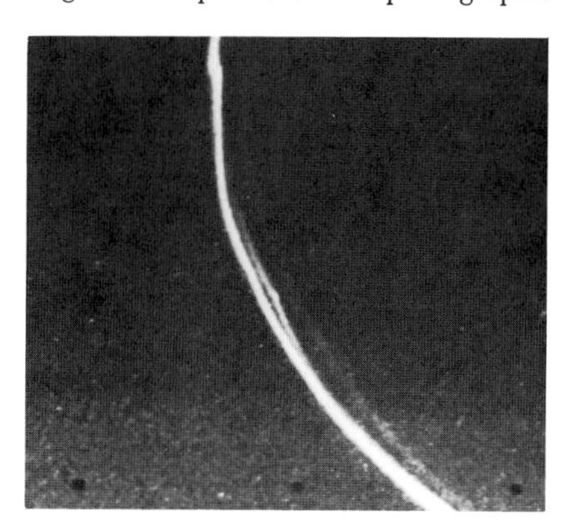

Figure 4.11
The multistranded F ring of Saturn as recorded by Voyager 1. (NASA photograph)

In spite of some difficulties with our scientific instruments during the final phases of the Voyager 2 flyby of Saturn, the encounters were successful beyond our most optimistic expectations. Within the brief period of a few weeks at Jupiter, and again at Saturn, more knowledge was accumulated than had been obtained during the entire span of human learning. Ahead lie Uranus and Neptune, and who can doubt but that new, enchanting worlds and still more exciting discoveries await us?

5

The New Sun

Randolph H. Levine

How can I say that the sun is new? Shall I claim, as did some ancient cultures, that the sun is periodically consumed by some deity or beast and then recreated? Or perhaps it sets into the ocean (the Earth being both flat and vast) and is extinguished at the end of each day. Obviously, the sun is not new in any of these ways. What is new about the sun is our understanding of its physical processes as revealed by relatively recent observations from both space and the ground and our realization of the extent of its influence on the entire region around it in space. This understanding is so different from the accepted explanations of 25 or even 10 years ago that today students are indeed confronted with a new sun.

The sun is a remarkable arena for several areas of science. Like a well-equipped playground, it provides activities that exercise a wide variety of developing talents. Astronomers and physicists study the sun because it is unique both as a star and as a physical laboratory. But the sun is also of concern to atmospheric scientists (meteorologists and climatologists) as well as almost the full range of space scientists. The diversity of causes and effects demonstrated on the sun makes it difficult to give a comprehensive view of it in any brief presentation. Yet, despite this diversity, there is one feature that unites the study of the sun both physically and conceptually: the solar magnetic field, which extends from inside the sun in all directions out to (at least) a distance equal to that of the orbit of Uranus. This entire envelope is called the heliosphere. The solar magnetic field plays a direct role in all parts of the heliosphere. Phenomena as diverse as solar flares, the solar wind, the magnetospheres of the planets, auroras, sunspots, the solar cycle, cosmic rays, and the solar corona cannot be understood without reference to the sun's magnetic field.

The oscillating sun

For almost 20 years it has been known that the photosphere, the visible surface of the sun, oscillates. If one looks at almost any point on the photosphere, it can be seen to rise and fall with a period of about 5 minutes, with the motion apparently organized in "granulation cells" several hundred kilometers across. Granulation cells can be seen as a fairly irregular, honeycomblike pattern in the photosphere. This was originally thought to be a localized phenomenon, much like ocean waves on the surface of the Earth. The 5-minute oscillation at any one point on the sun was not thought to be correlated significantly with the oscillation at any other distant point. However, within the past 5 years it has been confirmed that the 5-minute oscillation is a global phenomenon, the result of vibrations of the entire sun. Like the ringing of fine crystal, the sun as a whole oscillates in a precise pattern of frequencies and spatial scales.

Figure 5.1 is a graphic representation of the sun's global oscillations. The horizontal axis represents frequency (the reciprocal of the oscillation period), and the vertical axis is the wave number (the reciprocal of wavelength). The intensity displayed is greatest for those combinations of spatial scale and temporal period where the oscillations are strongest. One might guess that such a plot would show rather random (or at least smoothly varying) intensity, but the remarkable conclusion is that only very specific combinations of frequency and size oscillations exist on the sun. It is truly a very fine-tuned object. Perhaps equally remarkable is the fact that the bright "ridges" in figure 5.1 lie almost exactly along the lines predicted by theoretical studies of the sun's interior structure.

The important thing about these global oscillations is that they are not a property of just the surface, but of the entire body of the sun. The interior oscillates as well as the surface. Each type or mode of vibration is strongest in a different part of the sun, depending on its frequency and spatial scale. By studying the modes that are strongest at different depths, it is possible, in principle, to get an indication of the sun's interior circulation, because the rotation at different depths slightly modifies the properties of the oscillations. (For example, the rotation of the sun as a whole causes the symmetry of the pattern in figure 5.1 to be slightly skewed with respect to the axes. Nonsolid or differential rotation of the sun causes a different shift in different modes of vibration.) We should be able to discover whether the sun is rotating faster inside than at its surface by studying the oscillations that are strongest deep inside. Until now, this has been possible only

Figure 5.1
Graphical representation of the sun's global oscillations. The horizontal axis represents frequency, which is the reciprocal of the oscillation period; the vertical axis represents the wave number, which is the reciprocal of the wavelength. The intensity displayed is proportional to the strength of oscillations with different frequencies and wave numbers. The dark horizontal line is near wave number zero (i.e., very long wavelengths). Positive and negative wave numbers, respectively, represent waves traveling east and west on the sun. (There is no true zero of frequency. The pattern has been reflected about the horizontal axis, and all frequencies are positive.) The slight tilt of the pattern is due to the rotation of the sun. (Courtesy of J. Harvey, T. Duvall, and E. Rhodes, Kitt Peak National Observatory)

within a fairly thin layer near the surface, but the results are tantalizing. The interior does seem to be rotating faster (at least within the outer, convective part of the sun).

The greatest limitation to this type of study is the need to observe many, many wave periods in succession to obtain a good statistical picture of the exact frequencies and sizes of the waves. (Figure 5.1 is based on several separate days' observations.) The reason we cannot study the deeper wave modes is that we cannot look for longer periods of time—night interferes with solar astronomy. However, this particular problem can be partially overcome by choosing a suitable observing site. A French-American team of astronomers did this in late 1979 and early 1980 by going to the South Pole. The South Pole during the local summer is an excellent site for these observations of the sun. The climate is that of a desert, with very little precipitation, and the altitude is 11,000 feet. Astronomically speaking, it is high and dry. There are other astronomical advantages as well. No correction needs to be made for the Earth's rotation, and the sun traverses the full circumference of the horizon at the same altitude in the sky during the course of the day. Thus, the simplest telescope and detector arrangements are the most practical, which is fortunate because of the site's peculiar disadvantages: a rather short observing season and temperatures averaging near −40°F.

During their South Pole stay, the French and American scientists obtained up to 120 hours' continuous observations of the sun. Their telescope had no spatial resolution at all; it simply measured the net motion of the solar surface averaged over the entire disk of the sun. Even so, oscillations were not only detected, but were separated into a myriad of frequencies clustering near 5 minutes. These are some of the discrete frequencies at which motion detectable on the scale of the sun itself exists. Such techniques for observing solar vibrations thus hold great promise as tools for probing the sun.

Magnetic fields

Magnetic fields determine the motion of electrically charged particles, which are abundant throughout the cosmos. A magnetic field is a conceptual tool that assigns to each point in space a direction and a magnitude, or strength. Knowing the direction and the strength of the magnetic field at a point in space allows us to calculate the motion of an electrically charged particle at that point. You cannot see a magnetic field, or touch it. It is one of those philosophical necessities that help us understand (or think we understand) action at a distance: why two magnets are attracted to one another even though there is nothing between them that can be "pulled."

A convenient way of describing magnetic fields is the concept of magnetic field lines. If we draw smooth curves that go in the direction of the magnetic field at each point in space, the result is a pattern of magnetic force, or a set of magnetic field lines. Another way to understand magnetic field lines is to imagine moving a small compass through a magnetic field, always traveling in the direction in which the compass needle points. The compass path is always along a magnetic field line; start the compass at a new position and it will follow a different field line.

Magnetic field lines are not observable, and in some situations they cannot even be defined. But they are very useful as a concept, partly because there is an excellent analogy for many of their properties: rubber bands. They can pull and be pulled; they carry tension; and they even knot up if you twist them too much. This analogy is also helpful in understanding how they determine the paths of electrically

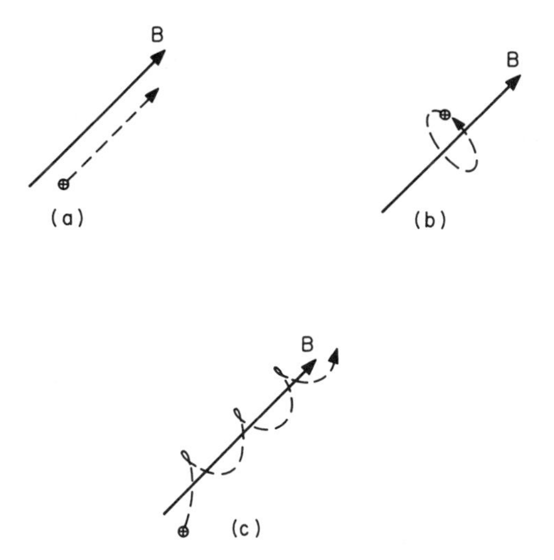

Figure 5.2
Schematic illustration of the motion of an electrically charged particle in a magnetic field. (a) Particle moving along local magnetic field direction; (b) particle moving perpendicular to local field direction; (c) particle moving at angle to local field direction.

charged particles. In the sun this is extremely important, because beyond a few thousand kilometers above the photosphere the outer atmosphere is so hot that atoms are stripped of electrons and all the particles become electrically charged. Thus, the magnetic field in the corona (and beyond in the interplanetary medium) determines the motion of this material.

How do electrically charged particles move in a magnetic field? The rule is simple: Charged particles moving *along* a field line follow the direction of the line, but charged particles trying to move *across* a field line move in circles around the line. So, in general, a charged particle whose motion is at some angle to the field line will move in a circle that slides along that line. In other words, it follows a spiral path. The radius of the spiral depends on both the velocity of the particle and the strength of the magnetic field. Larger velocities lead to larger spirals, as do weaker magnetic fields. In the solar corona, the average velocity is determined by the temperature, which is over 1 million °C, and that velocity is about 200 kilometers per second. The resulting radius for the spiral orbit of charged particles, given a typical coronal magnetic-field strength, is a few centimeters or less. This is a very important deduction, for it allows us to predict the structure of the corona. The material in the corona, being electrically charged, moves

Figure 5.3
The sun's corona as photographed from Kenya during the total eclipse of June 30, 1973. Most of the brightness is due to the scattering of photospheric light by electrons in the corona. (Courtesy of High Altitude Observatory, National Center for Atmospheric Research)

in very small spirals (radius of orbit about a centimeter or less, versus the radius of the sun, which is 70 billion centimeters). The particles are, in a sense, "tied" to a magnetic field line and move freely only *along* a magnetic line of force. This means that the solar corona should be very highly structured by the magnetic field, with coronal material distributed along field lines. Since these same particles emit the radiation we detect on or near the Earth, we can look for this structure by taking pictures of the corona at eclipses. Most of the structure seen in this way is on larger scales, but much of it does have the characteristic archlike shape of magnetic field lines connecting two points on the sun. (Some of the structure has a more fanlike shape.)

Eclipse pictures are impressive, and they do give evidence for the role of the magnetic field in the corona, but they show us only the corona at the limb of the sun. Because the corona is transparent, we are often looking through several structures at once, thus blurring the details of each. To better separate and define structures along our line of sight, it is necessary to look at the corona directly above the disk of the sun. An excellent way to do this is by observing the short-wavelength (approximately 10–900 Ångstroms) radiation, such as x-

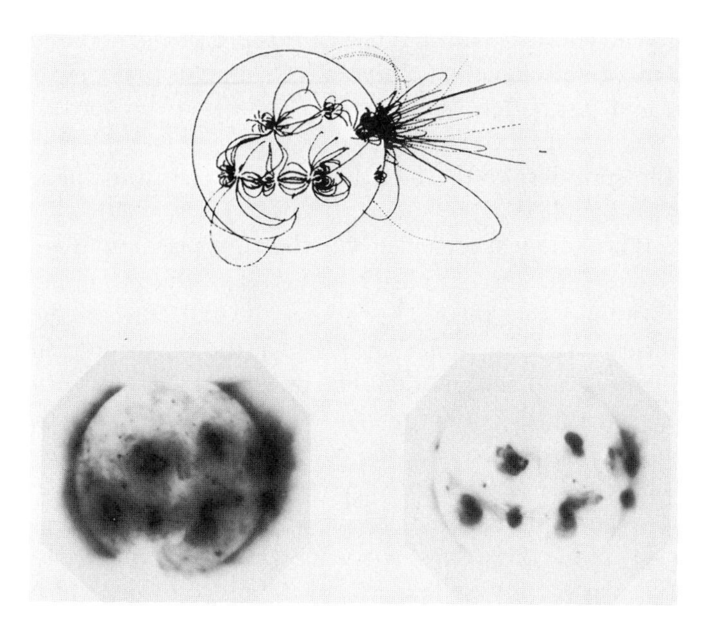

Figure 5.4
(Top) Theoretical diagram of magnetic field lines in the solar corona, based on observations of the magnetic field at the sun's surface. (Bottom) Images of coronal x-ray emission (two different exposures) at a time corresponding to that of the diagram above. Comparison of the magnetic-field calculation with the observed structure of the corona suggests that the topology of the corona is largely determined by its magnetic field.

ray or extreme ultraviolet radiation, emitted by the high-temperature solar material in the corona and not by the lower levels of the atmosphere such as the photosphere. An ultraviolet or x-ray picture of the corona—which has to be taken from space because the Earth's atmosphere absorbs ultraviolet and x radiation—reveals a rich view of the corona. All coronal structure observed in this way is directly attributable to magnetic fields. The hot corona resolves into a series of large and small loops of gas, each confined by "bundles" of magnetic field lines (sometimes called magnetic flux tubes). In short, the entire topology of the corona is ordered by the magnetic field.

But what makes the corona so hot? The heating of the solar corona has been one of the classic unsolved problems of astrophysics. It still remains both classic and unsolved, but now the formulation of the problem is very different from what it was two decades ago. Heating the corona now means heating magnetic flux tubes. It means supplying nonthermal energy throughout individual magnetic structures in order to maintain the plasma, or gas of charged particles, at the high temperatures observed. There are three stages to this process: generation of nonthermal energy, propagation of the energy to appropriate parts of the corona, and dissipation of the energy locally in the corona. None of these stages can be considered well understood in any detail anywhere on the sun, despite the fact that many mechanisms have been proposed and studied.

This problem is now often approached with the same plasma-physics techniques used in research on controlled nuclear fusion. One promising avenue is the possibility that the energy source for the corona is the magnetic field itself. We know, for example, that the magnetic field is continually changing and being replenished by the turbulent motions just beneath the sun's surface. Energy can be stored in the magnetic field and might be extracted to heat the coronal plasma over a long time—perhaps in ways similar to the very rapid release of energy in solar flares (the large explosions that occur on the sun from time to time). If this or any of the other suggestions for coronal heating mechanisms is correct, plasma physics tells us that the processes by which the energy is propagated and dissipated in the corona are likely to occur on very small spatial scales. For example, it is possible that strong currents flow in thin sheets only a few kilometers or at most tens of kilometers thick. Compare this scale with the size of a typical coronal loop (100,000 kilometers) or with the radius of the sun (700,000 kilometers).

The heliosphere

The solar magnetic field has been described as if its topology was fixed independent of the coronal material. To a large extent that is the case. The field is "rooted" in the photosphere, which is largely electrically neutral. The evolution of the photospheric portion of the solar magnetic field is determined by the same interaction between interior turbulence and solar rotation that produces the 11-year solar cycle. The motion of the sun's interior pulls at the magnetic field, stretching and distorting it as if it were a giant rubber band. When the field comes to the solar surface, we are often able to see the results of this push-pull process as sunspots.

The regularity of the sunspot cycle is well known but not entirely understood. Some of the important characteristics of the solar cycle are the variation in the number of sunspots, their tendency to appear in bipolar groups, the migration of the zones of sunspot appearance toward the equator during each cycle, and the magnetic polarity of sunspot groups. All these features are merely the surface aspects of the interaction among convection, rotation, and the magnetic field in the interior, and this is one reason why knowing the rotational properties of the layers beneath the surface will be important to our understanding more about the sun.

The solar magnetic field extends outward into the corona, where the gas density is lower and the gas is ionized. Such a gas is called a plasma. The ultimate source of the magnetic field is moving electrical charges, or currents, including those charged particles whose motion is controlled by the field. A competition exists between the field and the plasma to see which can push the other around. If enough plasma flows in a way that results in a current, the magnetic field generated by this new current can be as large as the field that guided the flow in the first place. This magnetic field is added to the original field to produce a total field that can be quite different from the original one.

The study of this complicated feedback process, called magnetohydrodynamics, is simply the study of flows in the presence of magnetic fields. When the fluid is a coronal-type plasma, a tension is set up: The field tends to push the fluid around, while at the same time the resulting fluid motions (and possibly other effects) produce currents that tend to push the field around. In the low corona, magnetism wins and the plasma is field-dominated; that is, the gas stays in place within magnetic loops and does not change the ambient magnetic structure very much. However, in the far outer corona and the interplanetary medium, the plasma is fluid-dominated; that is, the field lines are dragged with the flow of the plasma.

A good example of the balance between field and fluid in the corona is the existence of transient disturbances that move through the region (figure 5.5). Given sufficient impetus, such as a solar flare or other eruption, what looks like a magnetic bubble moves outward with explosive force. In the wake of these coronal transients, which are probably not so much bubbles as flattened arches, the coronal magnetic field can be significantly distorted over large regions for several hours. In most cases, the field then reestablishes its dominance of the lower corona and returns to a topology similar to that before the disturbance. Coronal transients probably occur about once per day.

In between the two extremes of a field-dominated low corona and a fluid-dominated outer corona, in the region between about two and ten solar radii, a very complicated and dynamic balance is set up. This is the region where the solar wind is born. The solar wind is a flow of plasma extending from near the sun to far out in the heliosphere. Such a flow can exist near any star with a hot outer atmosphere, with some of the energy lost by the hot plasma through conduction to

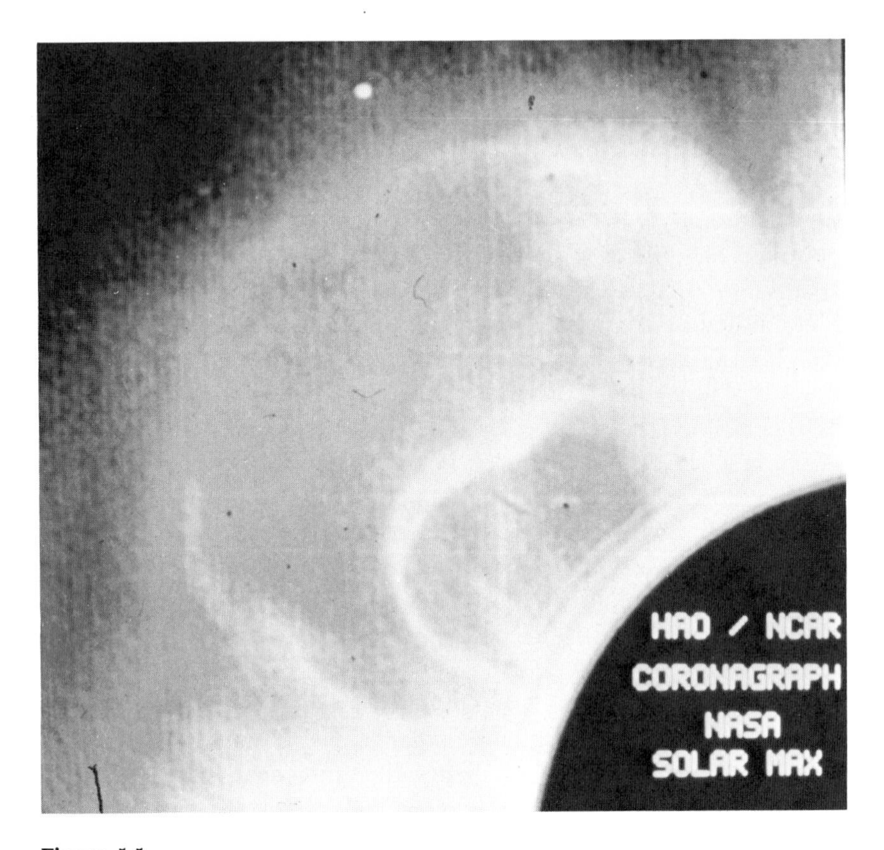

Figure 5.5
The sun's corona as photographed from the Solar Maximum Mission Satellite. The dark disk is part of the coronagraph, which blocks light from the solar surface. The disk's edge is about 0.75 solar radius from the limb of the sun. The large "bubble" is a transient coronal disturbance moving outward at approximately 500 kilometers per second. (Courtesy of High Altitude Observatory, National Center for Atmospheric Research)

cooler outer regions used to power the flow. However, the conversion of heat energy alone cannot account for the solar wind. It flows too fast for heat loss to explain both its speed and the amount of material involved. Some other processes must accelerate the plasma in the solar wind. At present, interest is centered on the possibility that wavelike motions of the plasma and the magnetic field carry energy and momentum to the solar-wind plasma. Such waves could be generated in the turbulent layers of the sun's interior.

Where the solar wind originates on the sun was once a great mystery. However, we can now use what we know about plasmas and magnetic fields to predict its source. We know that the particles have to spiral along magnetic field lines. Further, we can see from x-ray and ultraviolet satellite images that the looplike structures occupying most of the corona are tied to the solar surface at both ends—they are "closed" field lines. No plasma will escape the sun along their path. However, some of the higher loops can be disrupted by sufficient plasma pressures. (Remember, a strong enough plasma flow can push around a weak enough magnetic field.) These loops are distorted and eventually stretched out far into the heliosphere by the solar wind—they are "open" field lines. This process is responsible for the structure of the outer corona and results in the long, thin streamers and fan-shaped features so obvious in eclipse photographs. In short, the solar wind stretches the magnetic field lines of the sun far from their original place and shape. The plasma (which still has to flow along the field lines, even if it has distorted them) can flow far into the heliosphere along the open magnetic pathways. Thus, the location of at least some sources of the solar wind should be near where the highest closed field lines are rooted. Such places might look like the part in a head of hair, with the open field lines sticking out like a persistent cowlick.

This deduction is strikingly true. Not only does much of the solar wind originate in open magnetic regions, but these places are much cooler and less dense than the rest of the corona. And because they are also less bright than the rest of the corona, they are known as coronal holes. From these regions flow the strongest solar winds, called high-speed solar-wind streams. The coronal holes and the high-speed streams can last for months and are especially intense during those years preceding a minimum of the solar cycle. In fact, their full significance was first realized during the operation of the Skylab space station shortly before the last minumum of solar activity in the early 1970s. During that part of the cycle when activity is increasing, coronal holes occur at higher latitudes on the sun and are smaller than the longer-lived holes seen during the declining part of the cycle. One reason coronal holes are important is that the high-speed solar-wind streams associated with them have a particularly strong effect on the Earth's magnetic field. Exactly how the high-speed streams and their variation through the solar cycle might affect Earth (and especially its climate) is an active area of study.

The contrast between holes and loops in the corona is striking. The temperatures, the densities, and the organized flows are different. Their x-ray and ultraviolet radiations are totally different, and their magnetic structures are complementary. They can be studied almost completely independently, except for two very important facts. First, both holes and loops are connected magnetically to the photosphere and cannot be distinguished there. If one looks at the magnetic field in the photosphere, or at a particular wavelength of photospheric radiation, there is no way to tell a region under a coronal hole from one beneath a closed loop. The second thing common to holes and loops is that it takes about the same amount of energy input (per unit area) to keep coronal loops hot as to keep the solar wind flowing in coronal holes. This remarkable coincidence may not be a coincidence at all. Many scientists argue that such disparate types of physical structures could have the same overall energy budget only if the ultimate source of the energy is the same. The commonality of the photosphere and that of the turbulent convection zone beneath the photosphere strengthen this argument. Although the results are very different, the energy may come from the same source. In this view, the striking difference between the physical properties of coronal holes and coronal loops depends critically on different local processes, rather than on different energy sources.

As already mentioned, the solar wind distorts the outer corona, pulling part of the coronal magnetic field into interplanetary space. Out to a distance of about 3 solar radii from the center of the sun the flow is essentially radial, but at about 10 solar radii the weakening field lines can no longer resist the rotation of the sun. They begin to wind up, dragged along by solar rotation. Because the solar wind is continually pushing the field lines outward, however, the total effect is the creation of a spiral-shaped interplanetary magnetic field (figure

Figure 5.6
X-ray photographs of a coronal hole taken in 1973 and 1974. The successive images, taken one solar rotation (about 27 days) apart, show the evolution of a large coronal hole. Coronal holes can last for several months and represent magnetic configurations that connect the sun to interplanetary space. (Courtesy of American Science and Engineering)

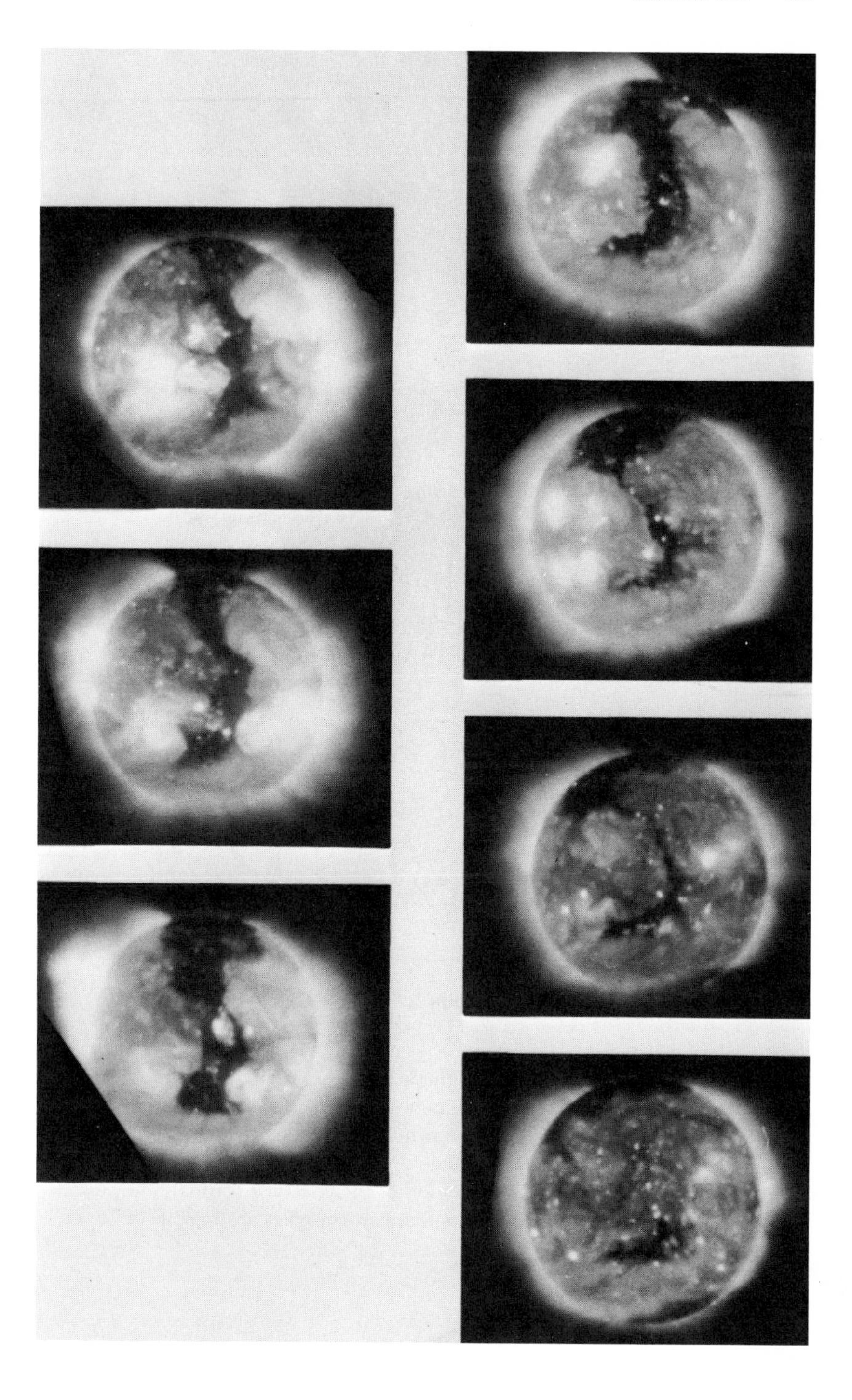

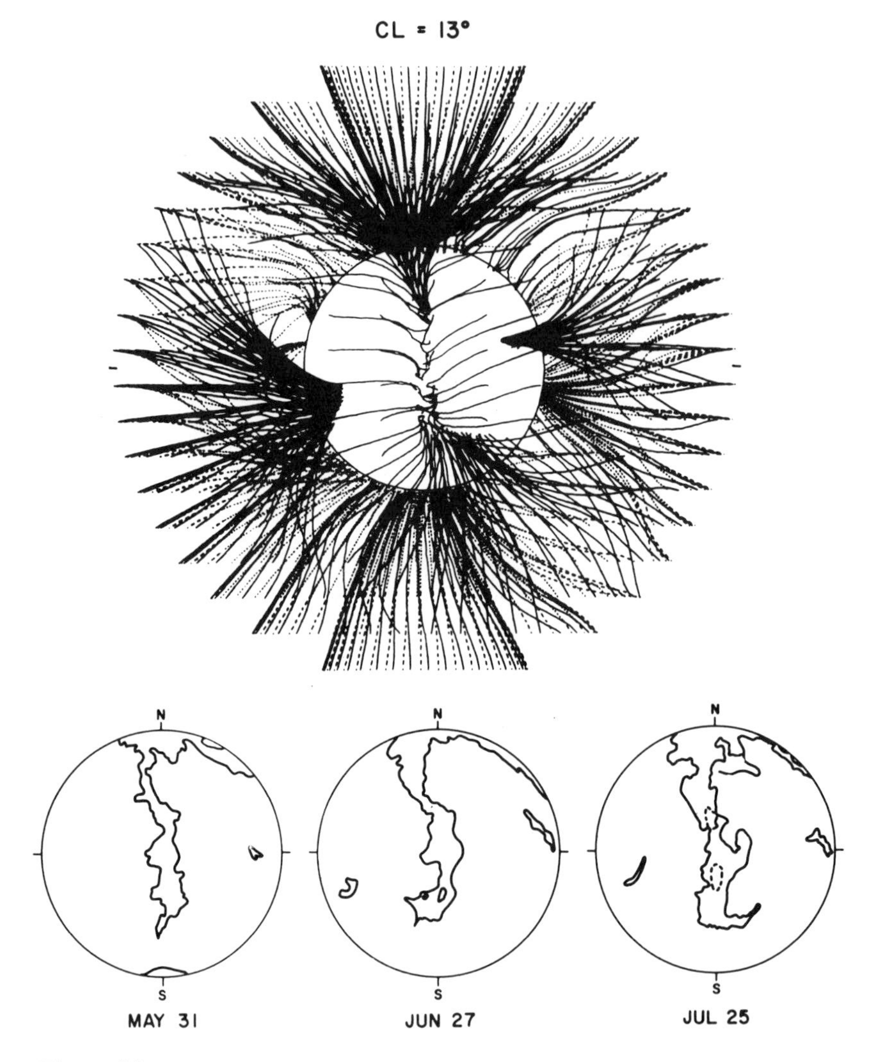

Figure 5.7
(Top) Theoretical diagram of solar magnetic field lines that connect to interplanetary space, based on the observed magnetic field at the surface. (Bottom) Tracings of the boundary of the coronal hole shown in figure 5.6 on three successive rotations of the sun. The central tracing (June 27, 1973) corresponds to the time of the theoretical diagram above. Comparison of the observed coronal-hole boundaries with the theoretical diagram shows that coronal holes are regions of "open" magnetic field lines.

Figure 5.8
Artist's conception of the interplanetary medium out to the orbit of Jupiter. The surface depicted is called the interplanetary neutral sheet; it is the surface that divides magnetic field lines pointing toward the sun from those pointing away. This surface is not visible, but it is shown here to emphasize the spiral structure of the interplanetary magnetic field and the wavy shape of the neutral sheet. (Courtesy of J. Wilcox, Stanford University)

5.8). The entire pattern rotates with the sun, like a giant pinwheel. One complete rotation takes about 27 days, as viewed from Earth. Long-lived features within the interplanetary field pass Earth every 27 days and are responsible for recurrent phenomena in our atmosphere. This spiral pattern is the basic magnetic structure embedded in interplanetary space.

Perhaps because it cannot be photographed or otherwise detected in a visually simple way, the interplanetary magnetic field has not been popularized as an important aspect of the solar system. Yet its potential significance to the planets and satellites is immense. The interplanetary magnetic field represents a direct connection to the sun, determines the path of charged solar particles in the heliosphere, and exhibits significant variability. The variability is important because a magnetized body in the solar system (Earth, for example) is subjected to a changing ambient magnetic field.

The steady flow of the solar wind past the Earth pushes the interplanetary field against the Earth's magnetic field, distorting it and creating a standing shock wave (a sort of constant magnetic analog of a sonic boom) on the sunward side and a long magnetic "tail" on the night side. The entire volume within the interaction region is the Earth's magnetosphere. Other planets also have magnetospheres—most notably Jupiter, which has a strong magnetic field of its own and a very active magnetosphere. The standing shock on the sunward edge of the Earth's magnetosphere can move, and can even distort or briefly tear, in response to variations of the interplanetary magnetic field. There is also some highly controversial evidence that, when the interplanetary magnetic field changes sign (that is, when the "neutral sheet" passes the Earth), unknown processes favor the development of low-pressure systems in the northern hemisphere. If this proves true, then the structure of the interplanetary magnetic field, which is in turn governed by the interaction of the solar magnetic field and the coronal plasma, has a significant impact on the short-term state of the Earth's atmosphere. In other words, it may affect the weather. Certainly this seems true on longer time scales. For example, it is thought that during one period of prolonged lack of magnetic activity in the seventeenth century (the so-called Maunder Minimum, when there were no sunspots for 75 years), the Earth's climate became slightly but significantly colder. The question of the sun's influence on the Earth's weather and climate is far from settled, but it is clear that the sun's magnetic extension into the heliosphere must be understood as part of any solution.

The expanding of the heliosphere must have a limit. At some point, the interstellar gas of our galaxy should provide sufficient pressure and the solar wind should slow down enough so that the solar magnetic field will expand no farther. The envelope thus created is the heliosphere, the region outside is part of the interstellar medium, and the transition zone is known as the heliopause. The position of the heliopause may even vary with the solar cycle, owing to systematic changes in the solar wind.

Theoretically, the heliopause ought to be about 20 astronomical units from the sun. (One A.U. is the mean distance from the Earth to the sun.) Twenty A.U. is also about the distance of the orbit of Uranus. However, the Pioneer 10 spacecraft, launched in 1972, is now over 25 A.U. from the sun and continues to detect the solar wind and the solar magnetic field. One reason for this greater-than-expected extent of the heliosphere may be that the solar wind detected by Pioneer 10 at 25 A.U. has not slowed down as much as had been expected, and thus can exert a greater push against the interstellar medium.

The spiral pattern of the interplanetary magnetic field is the vehicle for both the steady flow of the solar wind and the transient bursts of particles caused by solar flares. The very energetic particles created in solar flares are an interesting special case of charged-particle motion in the solar magnetic field. Enough energy is imparted to some particles accelerated in solar flares that their orbits cannot be closely tied to individual magnetic field lines. (The spiral orbits are simply too large.) Instead, they are confined within entire magnetic loop systems. Because of their large orbits and high velocities, these particles can travel great distances across the face of the sun, their paths guided by the large-scale magnetic field topology rather than by individual field lines. Some of the energetic particles remain trapped in the large loop systems of the corona for up to several days. Eventually, however, most of them encounter an open magnetic field region and their orbits leave the vicinity of the sun entirely. These energetic particles travel along the interplanetary field, and some may reach the Earth. If they penetrate the shock wave marking the boundary of the magnetosphere, they can travel along the Earth's magnetic field into our atmosphere and produce auroras.

Many areas have been addressed in this chapter, and each one is usually the province of a specialist. But the sun is also much like a playground: Seen from a distance it appears a somewhat chaotic whole, but seen close up there is a remarkable precision and cohesion to the activity in each of its separate parts.

Reading

Eddy, J. A. A New Sun: The Solar Results from Skylab. NASA report SP-402. Washington, D.C.: Government Printing Office, 1979.

Eddy, J. A., ed. *The New Solar Physics.* Boulder, Colo.: Westview, 1978.

Frazier, K. *Our Turbulent Sun.* Englewood Cliffs, N.J.: Prentice-Hall, 1982.

Friedman, Herbert. *The Amazing Universe.* Washington, D.C.: National Geographic Society, 1975.

Gallant, R. A. *Our Universe.* Washington, D.C.: National Geographic Society, 1980.

6

The Ultraviolet Sky

Andrea K. Dupree

The ultraviolet region of the spectrum is another of the new windows opened onto the universe by space technology. The view through this window has presented us with many surprises and, even after careful study, many puzzles. The most striking discovery has been the vision of a dynamic, inhomogeneous universe characterized by activity on every scale, from stars to galaxies.

The ultraviolet region of the spectrum also is well suited to the answering of many critical astronomical questions, because some cosmic elements and materials can be "seen" only at these wavelengths. These elements are among the most abundant and hence most significant in determining the character of the universe and its contents. Moreover, the hot material that forms stars and parts of the interstellar medium radiates copiously in the ultraviolet spectral region, and most of the phenomena related to this radiation cannot be inferred from the optical region alone. On an even more direct level, ultraviolet observations of stars that show features similar to those of the sun, but on much larger scales, confront our understanding of our own star and its relationship to the terrestrial environment.

Scientists realized decades ago that observing ultraviolet radiation from cosmic sources would require telescopes to be located above the absorbing layers of the Earth's atmosphere—especially the ozone layer, which shields life from the harmful effects of solar ultraviolet radiation. Early in this century, Theodore Lyman of Harvard University went to New Hampshire to see if the altitude of the mountains was sufficient to allow observations of the sun in the ultraviolet range. Alas, he was not successful. Neither were the early balloon experiments of the 1930s. After World War II, a few early rocket experiments reached altitudes high enough to obtain ultraviolet spectra of the sun. However, only since the advent of the space age, with Earth-orbiting satellites complementing scientific rockets and high-altitude balloons, have we been able to explore fully the ultraviolet region of the spectrum.

The first series of satellites sponsored by the National Aeronautics and Space Administration, the Orbiting Astronomical Observatories, laid the groundwork with a series of photometric and spectroscopic instruments constructed by the Smithsonian Astrophysical Observatory, Princeton University, and the University of Wisconsin. The last satellite in that series, OAO-3 (also called Copernicus), was launched in 1973. Copernicus was followed five years later by the International Ultraviolet Explorer satellite, which at the time of this writing is still operating. The OAOs and the IUE returned much information about stars, galaxies, and interstellar matter at ultraviolet wavelengths. In addition, NASA sponsored a series of experiments aboard the Orbiting Solar Observatory and Skylab satellites for ultraviolet observations of the sun, and the Apollo astronauts carried ultraviolet-sensitive cameras to the moon.

Physics of the ultraviolet

The ultraviolet spectral region offers an unusual opportunity to study hot objects in the universe and, under certain conditions, to detect cold material as well. Figure 6.1 shows the energy distribution of hypothetical objects—so-called black bodies—of different temperatures. (Real objects such as stars radiate energy in a qualitatively similar way.) The spectral distribution of energy varies depending on the temperature, with the radiation becoming more and more energetic (and the wavelengths shorter) as the temperature rises. Thus, relatively cool objects such as the sun, with temperatures below 6,000°K, have their maximum emission in the infared and visible regions of the spectrum, whereas objects at temperatures over 1,000,000°K tend to emit most of their radiation as x rays. In between these two bands is the ultraviolet region of the spectrum, which is optimal for objects with temperatures from 10,000°K to 1,000,000°K (including the interstellar medium and the stellar component of galaxies).

There are various ways to measure ultraviolet radiation. The most information results from a technique in which the radiation is not measured in its entirety but rather is dispersed into very small discrete intervals of energy or wavelength. This resolved radiation is termed a spectrum.

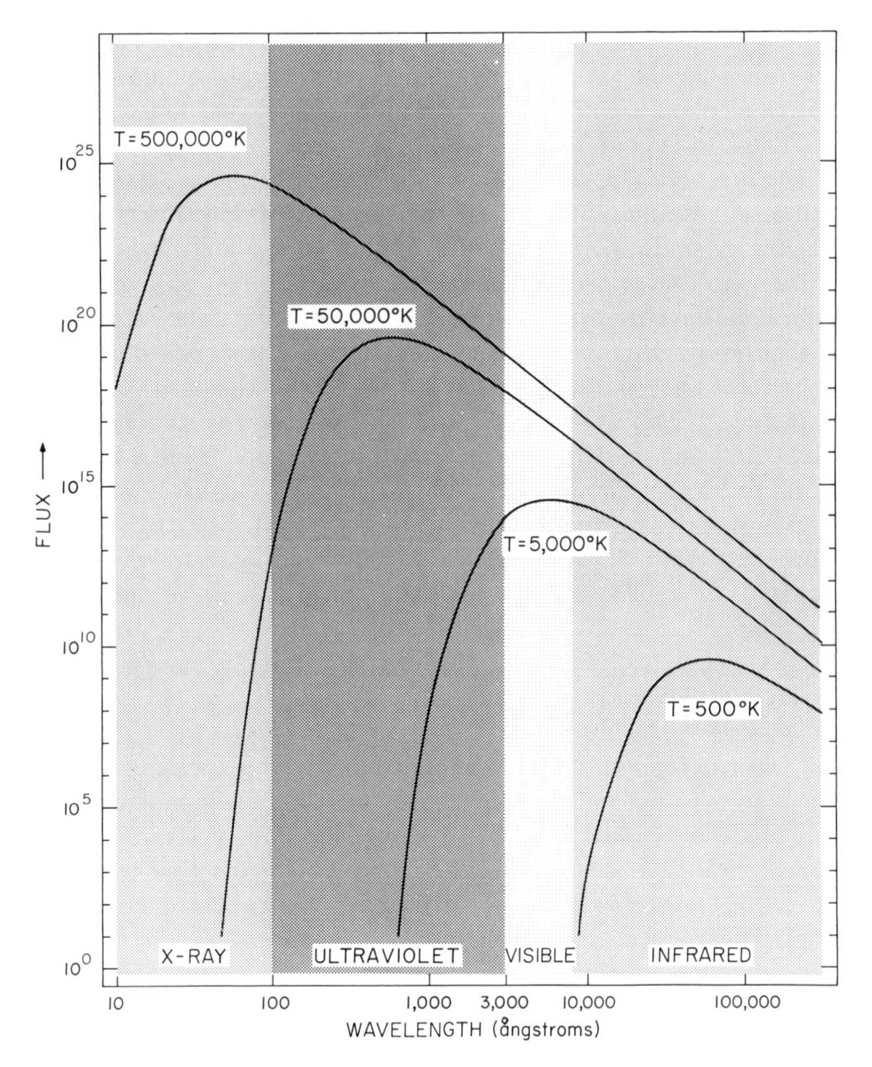

Figure 6.1
The energy distribution of hypothetical "black bodies" having different temperatures. The ultraviolet spectral region occurs between wavelengths of 100 and 3,000 Ångstroms, where objects having temperatures between about 5,000°K and about 500,000°K can be detected most easily. (Illustration by Beryl Langer)

A hot object, such as a compact star, will radiate a continuous spectrum, as shown in figure 6.2. The amplitude and the shape of the continuous spectrum indicate the temperature of the object and the physical mechanism producing the emission. However, the appearance of an astronomical spectrum also depends on the physical characteristics and orientation of any gas that may be present along the line of sight between the observer and the object. If the intervening gas is at a higher temperature, then a series of emission lines will arise from the gas to be superposed on the continuous spectrum. By contrast, if there is cold gas between the hot object and the observer, it will absorb the radiation from the hot object and again produce a distinctive pattern of absorption lines. These emission or absorption lines arise from atoms and ions residing in the gas; each atom and each ion has its own characteristic signature.

The most abundant atoms and molecules in the universe have their prominent spectral signatures in the ultraviolet region of the spectrum. These include the atoms of hydrogen, deuterium, helium, carbon, nitrogen, oxygen, and silicon and the hydrogen (H_2) molecule. Spectroscopy provides a powerful means of determining the characteristics of and the physical conditions in cosmic material that can be observed only at a distance. The presence of atoms or molecules in a spectrum indicates that the material must be cool, because molecules dissociate into their component atoms at high temperatures. Signatures of ions that are very highly excited (ionized) point to a very hot gas, because atoms lose electrons and form positively charged ions at high temperatures. The strength of these features indicates the abundance of a particular element or the excitation mechanism of the specific ion. Moreover, the position of the lines on the spectrum reveals the motion of the gas relative to the observer, with gas moving toward him producing a shift farther to the ultraviolet and gas moving away from him showing a shift to the red. A detailed measurement of the shape of a spectral feature can reveal the presence of turbulent motions in the gas and even the strength of any magnetic field that may be present.

A typical ultraviolet spectrum of a hot star taken with the International Ultraviolet Explorer satellite (figure 6.3) illustrates many of the above-mentioned characteristics. In the IUE's observing system, the ultraviolet spectrum is broken up into sections by a system of gratings and mirrors before it strikes the face of a detector similar to a television tube. The spectral information is read electronically from the tube in space and radioed back to Earth, where it is reconstructed to form an image. The streaks in this image indicate the continuum

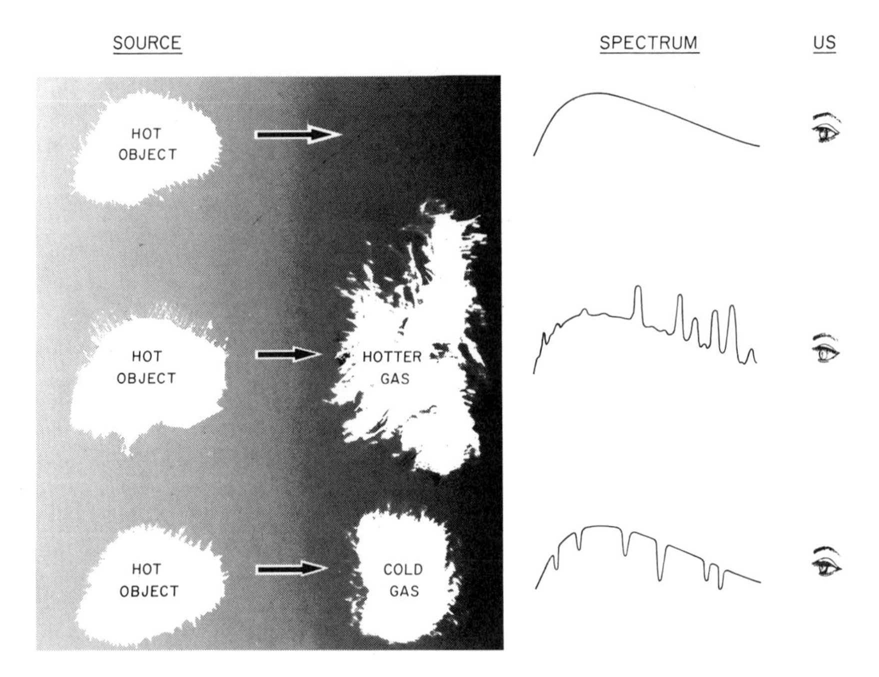

Figure 6.2
Schematic representation of the formation of a continuous spectrum (top), an emission line spectrum when hotter gas intercepts the line of sight (middle), and an absorption line spectrum when cold gas intervenes (bottom). (Illustration by Beryl Langer)

Figure 6.3
Ultraviolet spectrum of a hot star in the wavelength region 1,180–2,000 Ångstroms, taken with IUE. The dark line pattern marks the continuous spectrum. The spectrum shows the broad absorptions caused by the expanding atmosphere and the narrow absorptions arising from the intervening interstellar medium. The white dots are reference marks on the circular faceplate of the detector. (NASA/Center for Astrophysics photograph)

radiation from the star, and the shape of the spectrum approximates a black body with a temperature of about 35,000°K. This continuous radiation is crossed by narrow dark features (absorption lines arising from cool gas along the line of sight to the star) and by broad dark features (absorption lines marking the absorption in the star's own extended atmosphere). Each feature can be identified with a specific atom or ion to describe the compositions of the stellar atmosphere and the intervening gas. The extent of the broad absorption features indicates that the outer atmosphere of the star is expanding away from its surface at about 2,000 kilometers per second.

Solar-system objects

The highly successful probes to the other planets of the solar system produced an array of impressive optical images. With its higher spectral resolution and its ability to obtain observations over a longer period of time, the IUE satellite complemented these flyby missions to the planets and their satellites. The ultraviolet spectra led to the discovery of absorption features in planetary atmospheres, to the study of spatial and temporal variations in the atmospheres and the magnetospheres, and to the mapping of the surfaces. For example, Voyager photographs of Io, one of Jupiter's satellites, showed a mysterious white material distributed over much of the surface. There was no simple way to identify this material until ultraviolet spectra were obtained. The pattern of absorption lines discovered in these spectra corresponded to those that might be expected from frozen sulfur dioxide—a finding consistent with other evidence for volcanic activity on Io.

Earth is also visible in the ultraviolet range. From the airless moon, the Apollo 16 astronauts photographed the polar aurora and the tropical airglow enveloping the Earth (figure 6.4). The presence and the variation of these emissions are related to solar activity and give information about the structure of the Earth's magnetosphere.

The ultraviolet appearance of an object can be quite different from the visual. Comet Kohoutek sadly disappointed optical observers when it failed to produce a bright display in the night sky. However, when photographed in the ultraviolet range in a rocket experiment, Kohoutek was spectacular. In the principal ultraviolet line of hydrogen, the Lyman α line, the comet showed an enormous amount of hydrogen surrounding its head in a cloud 15 times the diameter of the sun. This hydrogen is believed to have resulted from the dissociation of frozen water vapor sublimed from the comet nucleus through solar heating. Similarly, the study of comets has been aided significantly by the IUE

Figure 6.4
Earth as photographed from the lunar surface during the Apollo 16 mission. The image, taken in the light of atomic oxygen (near 1,300 Ångstroms), shows the polar aurora and the tropical airglow belts. (Courtesy of G. Carruthers, Naval Research Laboratory)

spectroscopy, which has measured oxygen, OH, and other atoms and molecules in addition to hydrogen. Indeed, the images and spectra of many comets indicate that the compositions of these bodies are similar, suggesting a common origin.

Stars

Since the bulk of a hot star's radiation emerges in the ultraviolet range, this band is particularly useful for detecting those young, hot stars that burn nuclear fuel at such prodigious rates that they evolve quickly. In fact, the detection of this radiation enables us to map the regions of star formation in our galaxy and in others. Particularly striking evidence of this ability can be seen in figure 6.6, which shows the Large Magellanic Cloud in both visible and ultraviolet images. In optical light, this nearby galaxy appears as an amorphous collection of cool stars, but in the ultraviolet band hot objects dominate to reveal an extensive region of "chains" where stars have been recently formed.

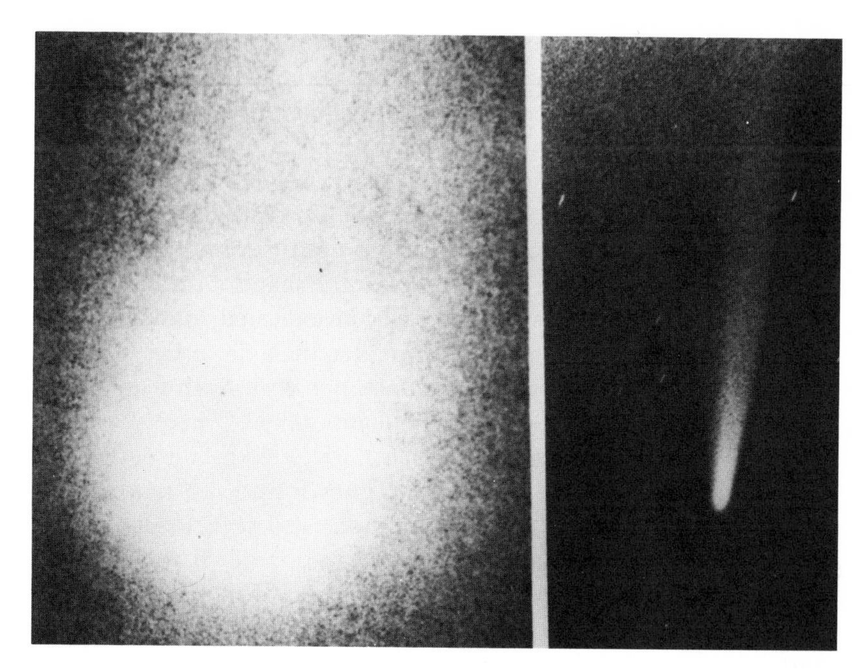

Figure 6.5
Comet Kohoutek, (left) as imaged in ultraviolet light of hydrogen (by Lyman α) and (right) as it appeared in visible light. (Courtesy of G. Carruthers, Naval Research Laboratory)

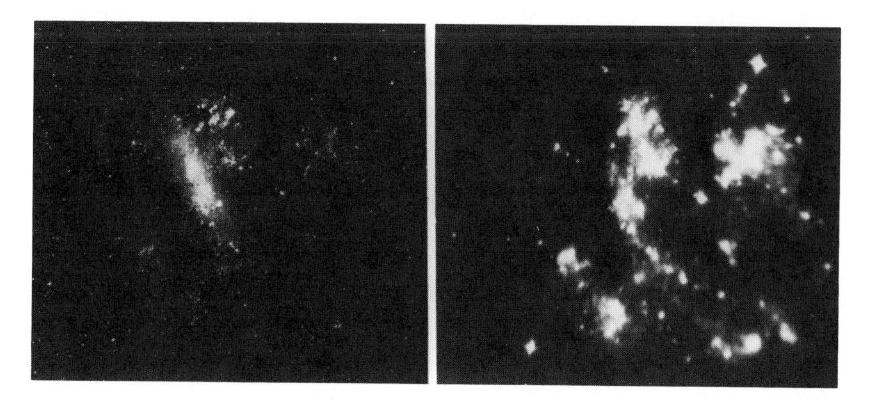

Figure 6.6
Two images of the Large Magellanic Cloud taken from the lunar surface during the Apollo 16 mission. The left image shows the optical appearance of the cloud; the right image was taken in the ultraviolet light centered near a wavelength of 1,450 Ångstroms. (Courtesy of G. Carruthers, Naval Research Laboratory)

A major result in stellar astrophysics has been the discovery that many stars have hot outer atmospheres (coronas) similar to that of the sun. Ultraviolet astronomy contributed to this discovery by revealing that material is driven off from young hot stars at extraordinarily high rates. Early rocket observations of several such stars produced spectral line patterns showing broad absorption, which indicated that gas was moving away from the stars at velocities up to several thousand kilometers per second (figure 6.7). Moreover, the shape of the line profiles for three-times-ionized carbon and silicon indicated mass loss rates from 10 times to 10 million times greater than the solar rate—the equivalent of a millionth of a solar mass per year. Such features are commonly observed in an expanding hot atmosphere, where absorption of the ultraviolet radiation from the star drives the atoms outward to produce a stellar wind. Subsequently, lines from even more highly excited ions of nitrogen and oxygen were also seen to display the distinctive signatures of an expanding atmosphere. It thus became increasingly difficult to explain the production of these ions without the presence of a hot atmosphere, or corona. Observations by the Einstein satellite confirmed that young hot stars have hot coronas because they emit x rays.

Some of these hot stars are known to have compact companions orbiting within their expanding atmospheres (figure 6.8). The companion may be a neutron star or, as may be the case with Cygnus X-1, a black hole. One of these binary systems, a hot supergiant linked to a neutron star dubbed Vela X-1, emits x rays at a prodigious rate, apparently as the result of accretion of the stellar wind upon the compact object. It is possible in the ultraviolet range to trace the presence and the effects of the hot x rays in the expanding stellar wind. As the neutron star passes in front of the supergiant star, the x rays heat and ionize a cavity in the stellar wind surrounding the neutron star. This means that the already three-times-ionized silicon near the neutron star becomes even more highly ionized and the absorption lines vanish. When the x-ray source moves behind the supergiant star, the wind is undisturbed by the x rays and the absorption profile returns to its usual form. Such variation in the absorption patterns allows study of the acceleration rate of the wind and offers clues to the structure and to the mechanism driving the mass loss.

Cooler stars such as our sun also show ultraviolet radiation, but their spectral images are much different because the hot gas of the atmosphere is now projected against a cooler continuum background. The solar chromosphere and corona have temperatures ranging from 10,000°K to 4,000,000°K, whereas the sun's visible surface (photosphere)

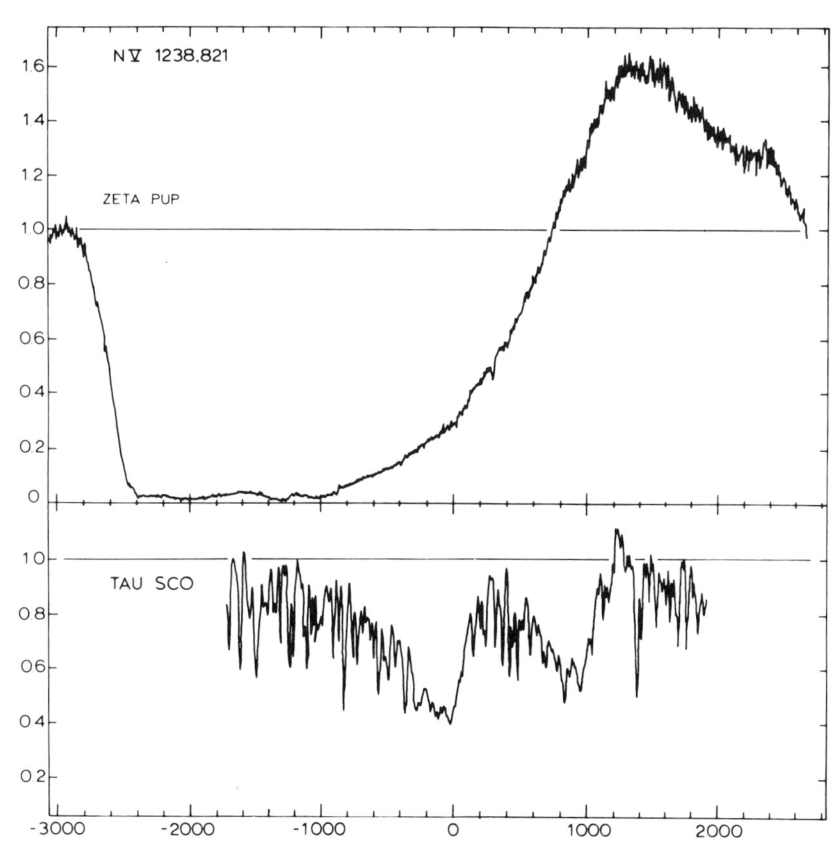

Figure 6.7
Line profiles of ionized nitrogen in two stars, Zeta Puppis and Tau Scorpii, showing the extended absorption expanding outward from the stars. These high-quality spectra are from the Copernicus satellite. (Courtesy of H.J.G.L.M. Lamers)

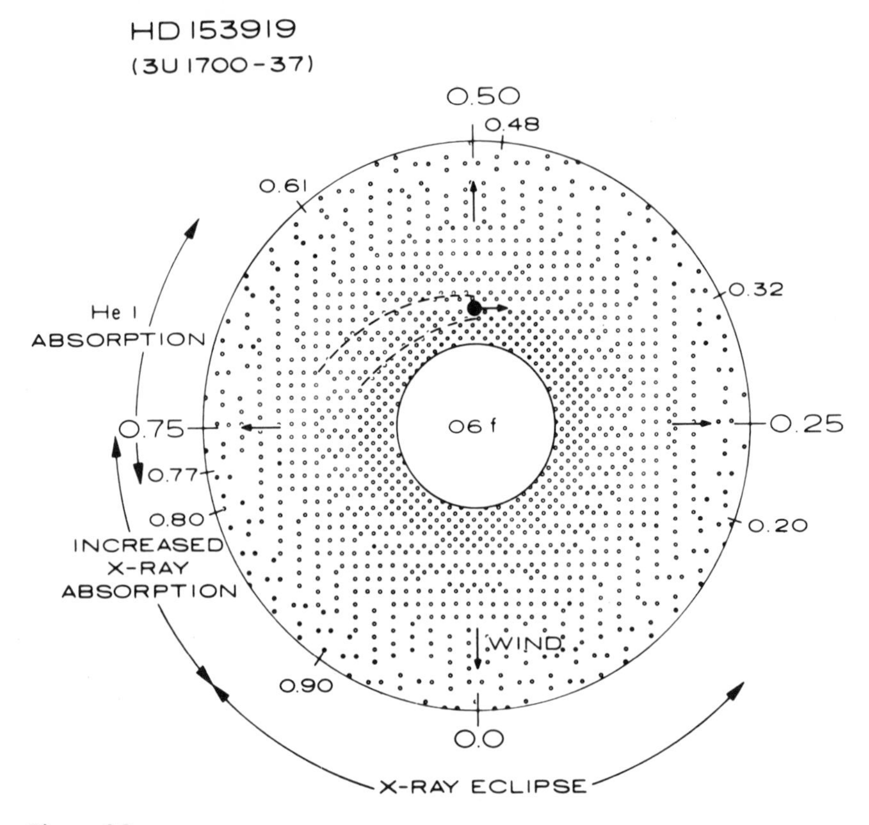

Figure 6.8
Schematic drawing of a compact object (filled circle) orbiting deep within the expanding wind of a massive hot star. Such a configuration produces strong x rays from accretion of the wind by the compact object. The numbers around the edge correspond to the phase of the orbit and indicate the orientation of the system; for instance, at phase 0.0 the compact object is eclipsed by the massive star, and at phase 0.5 the compact object is in full view.

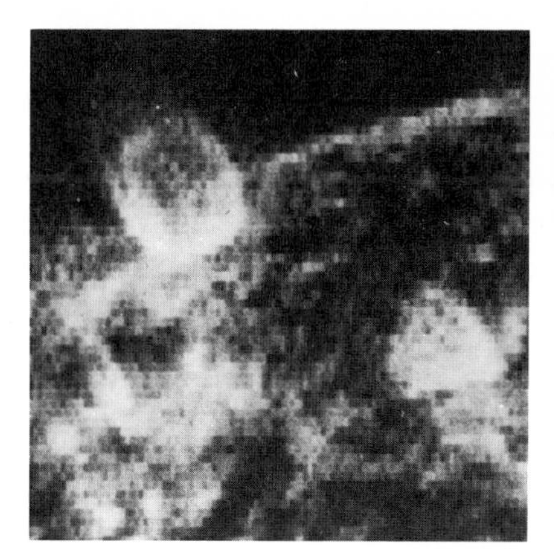
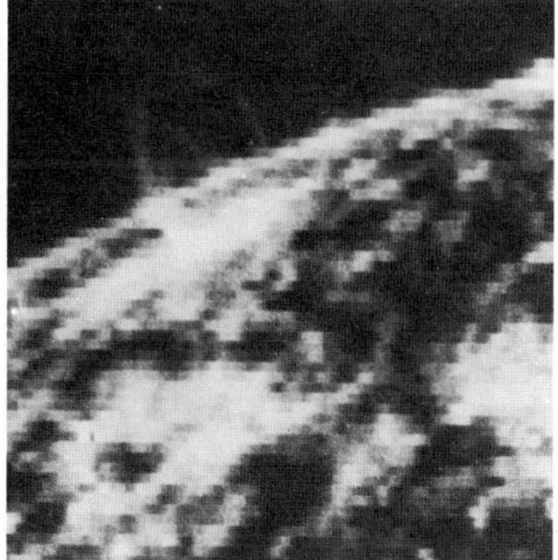

Figure 6.9
Active regions on the limb of the sun as photographed in ultraviolet emission lines. The structure of the atmosphere appears different in the various transitions. The inhomogeneous nature of the solar surface is apparent, and the importance of magnetic fields in controlling the atmospheric structure can be seen in the loop structures. (Harvard College Observatory photographs)

has a temperature of only about 6,000°K. Thus, an ultraviolet image of the sun shows emission lines rather than absorption lines, and that emission is greatly enhanced over regions of solar activity (figure 6.9).

From solar studies in which individual features on the surface can be resolved, it is known that hot, dense centers of activity and enhancement of ultraviolet emissions are associated with regions having a strong magnetic field. Where the magnetic field is closed, giant loop structures filled with hot plasma are formed. Where the magnetic field is open, the atmosphere ejects material into the solar wind that streams out into interplanetary space and by the Earth.

Apparently the same structures are also found on other stars—sometimes on a much larger scale. A particularly good example is the binary system Lambda Andromedae, which is composed of two giant stars, somewhat cooler and larger than the sun, in orbit around each other. The optical light from these stars fluctuates with a 54-day period. When the stars are faint, the colors of their visible surfaces indicate the presence of cool, dark, active regions ("star spots"). These spots diminish when the stars are bright. In ultraviolet light, corresponding spectroscopic signatures of active regions are also seen. The ultraviolet spectra show enhanced emission whenever the dark or "spotted" side of a star faces us, and decreased emission (and line profiles suggesting mass outflow) when the dark sides turn away or the spots diminish (figure 6.10). Much of this activity is on a substantially larger scale than that on the sun. These giant stars have diameters about ten times the solar diameter, and the "star spots" can occupy nearly 20 percent of the surface. On other stars, fully half the surface can be covered with spots. By contrast, on the sun, even the most extensive system of active regions covers much less than 1 percent of the surface. Clearly, extremes of activity can exist in stars that we could not guess from studying the sun alone, and these challenge our understanding of the nature of magnetic activity.

Even more spectacular examples of stellar winds and their variability can be found on supergiant stars. Stars of this class are close to the sun in temperature, but their diameters are about 200 times that of the sun. Supergiants are also losing mass at rates a million or more times the solar rate. Insight into the character of their stellar winds is provided by measurement of ultraviolet line profiles. The two spectral images of a supergiant star shown in figure 6.11 indicate a drastic change in the density of its wind over a period of 11 months. Through detailed study of such profiles it appears that a stellar wind is not a constant, smooth flow, but may well undergo changes in density and perhaps in velocity caused by the changing pattern of magnetic fields on the stellar surface.

Star clusters

Globular clusters, those massive aggregates of stars that surround our galaxy, attracted renewed attention after the discovery of intense x-ray sources near the centers of some clusters.

A globular cluster is a gravitationally bound group of a few hundred thousand stars within a spherical volume a few tens of light years across. In theory, the distribution of stars in a globular cluster should depend on the mass of the star. Thus, as the cluster evolves, heavy stars should sink to the center, and lighter stars should be propelled to the edge. This led naturally to a model in which a massive black hole located at the center was the source of x rays, but direct evidence for this was lacking. A second model proposed that the x rays were produced by mass accretion in a binary star system, but no binaries had ever been detected in the clusters.

The distribution of stars is particularly dense at the center of an x-ray cluster. It was difficult, if not impossible, to study individual stars at the very centers of such clusters. However, the distribution of ultraviolet radiation showed the shortest wavelengths to be more tightly concentrated across the cluster core in x-ray-emitting clusters than in

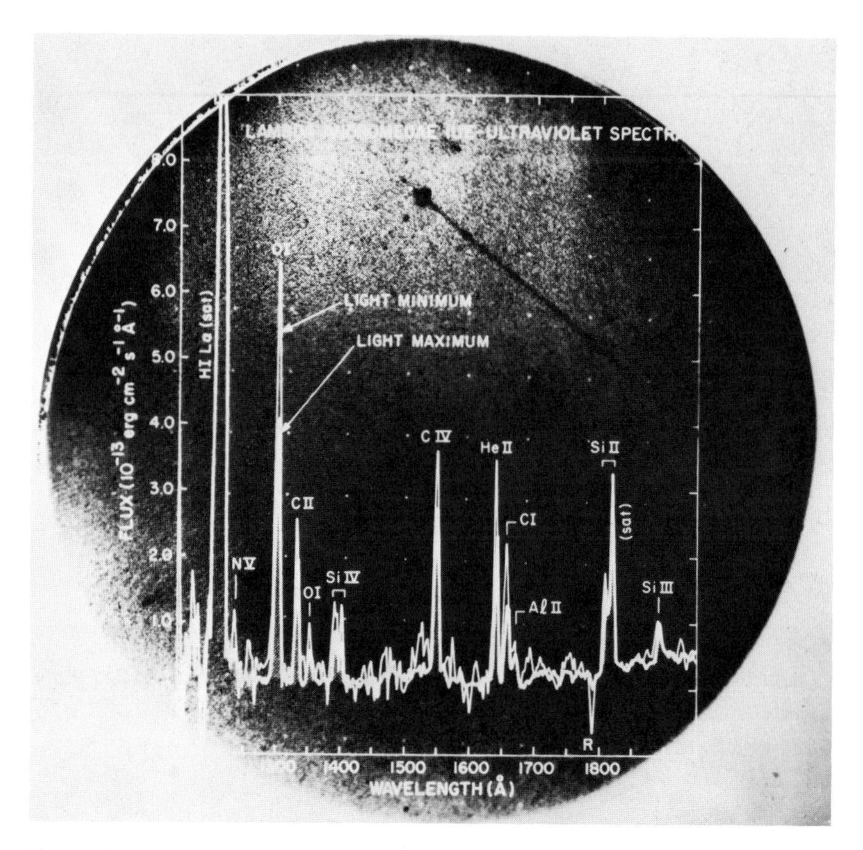

Figure 6.10
Ultraviolet spectrum of the active binary star Lambda Andromedae superposed on the image of the ultraviolet spectrum as recorded by the IUE spacecraft. The emission lines show fluctuations depending on the degree of activity of the star. Light minimum signals the presence of "star spots" associated with centers of activity and an enhanced level of ultraviolet emission; light maximum corresponds to an absence (or decrease) of the centers of activity.

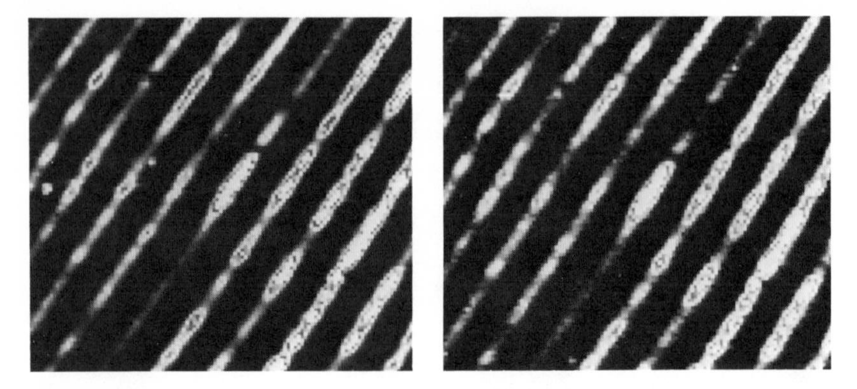

Figure 6.11
Two images of the ultraviolet spectrum of the cool supergiant star Alpha Aquarii. The central stripe corresponds to a line of Mg II at 2,800 Ångstroms and shows variable intensity corresponding to charges in the density of the wind.

Figure 6.12
A typical globular cluster, in Pegasus M15. (Hale Observatory photograph)

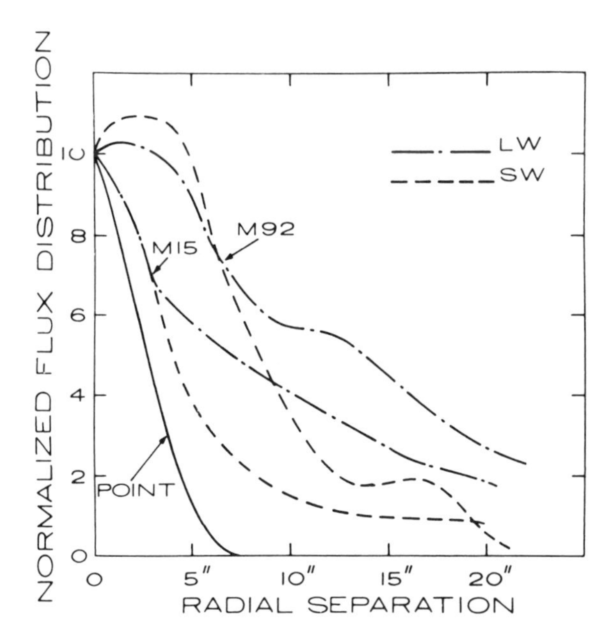

Figure 6.13
Tracings of the variation of ultraviolet continuum emission across the cores of two globular clusters, M15 and M92. The curves marked LW and SW correspond to "long" and "short" ultraviolet wavelengths. The curve marked "point" indicates the spatial resolution of the IUE telescope. Note that the distribution of the short-wavelength emissions from hot stars is narrower than that of the long-wavelength emissions from cooler stars.

normal clusters (figure 6.13). Since the short wavelengths arise from objects that are hotter, these observations indicated that hot stars were more likely to be found in the cluster cores than cool stars. It had been thought that the hot and cool stars in globular clusters had similar masses. This paradox may be explained if the hot stars are binary and hence more massive than the cool stars. Moreover, the presence of binaries may resolve the mystery of the x-ray sources in clusters without resort to the model of massive black holes.

The interstellar medium

Lying between the stars of our galaxy is a small but important component of gas and dust. The interstellar material comprises a few percent of the mass that is visible in stars, and the discovery of its character and extent has been one of the milestones of ultraviolet spectroscopy.

Much of the interstellar medium is visible only as absorption lines appearing against the light provided by some background object, and even small amounts of material lying along a line of sight can be detected. Since much of the interstellar gas arises from the winds streaming outward from the stars, it was logical to expect the elemental abundances of both to be similar. Astronomers were surprised to find that this was not true; instead, many elements were depleted in the interstellar medium relative to their values in the sun. For example, the ratio of calcium to hydrogen in some directions is a factor of 1,000 less than its solar value. It turns out that these depleted elements can form solids—grains of dust, such as irons, graphites, and silicates—that are not visible in ultraviolet spectra.

Molecules also can form in cool clouds (10–100°K), where the grains shield atoms from possible destruction by ultraviolet radiation. Enormous molecular clouds—some of the most massive objects in the galaxy—can develop, and it is in these dense clouds, often representing the remains of earlier stellar generations, that the cycle of star formation and evolution begins. The most abundant element in the universe is hydrogen, and its atoms combine to form the H_2 molecule, the most abundant molecule in the interstellar medium. Detection of this molecule in an ultraviolet spectrum in the direction toward a star (figure 6.14) was a fundamental discovery of modern astronomy.

Another element found in the interstellar medium is deuterium, an isotope of hydrogen containing a proton and a neutron in its nucleus. Although the deuterium is generally less than 1/50,000 as abundant as hydrogen in the universe, it can be detected by its absorption lines appearing against a background provided by a star. Deuterium is thought to have been formed in the very first moments of the universe—a time when densities and temperature were exceedingly high, thus allowing formation of certain light elements. Since there is no easy way to make deuterium except under such extreme conditions, the deuterium seen today may be primordial and may reflect the physical conditions that were operating in the "big bang." Those same conditions determine the total mass of the universe and its ultimate fate. The low abundance of deuterium implies an "open" universe, that is, one in which not enough material exists to stop the outward expansion and return the universe to a collapsed state. However, variations in the deuterium abundance appear from place to place in the solar neighborhood, and many are disturbingly larger than we can understand. Perhaps we do not yet fully comprehend all the ways this important isotope can be formed and destroyed.

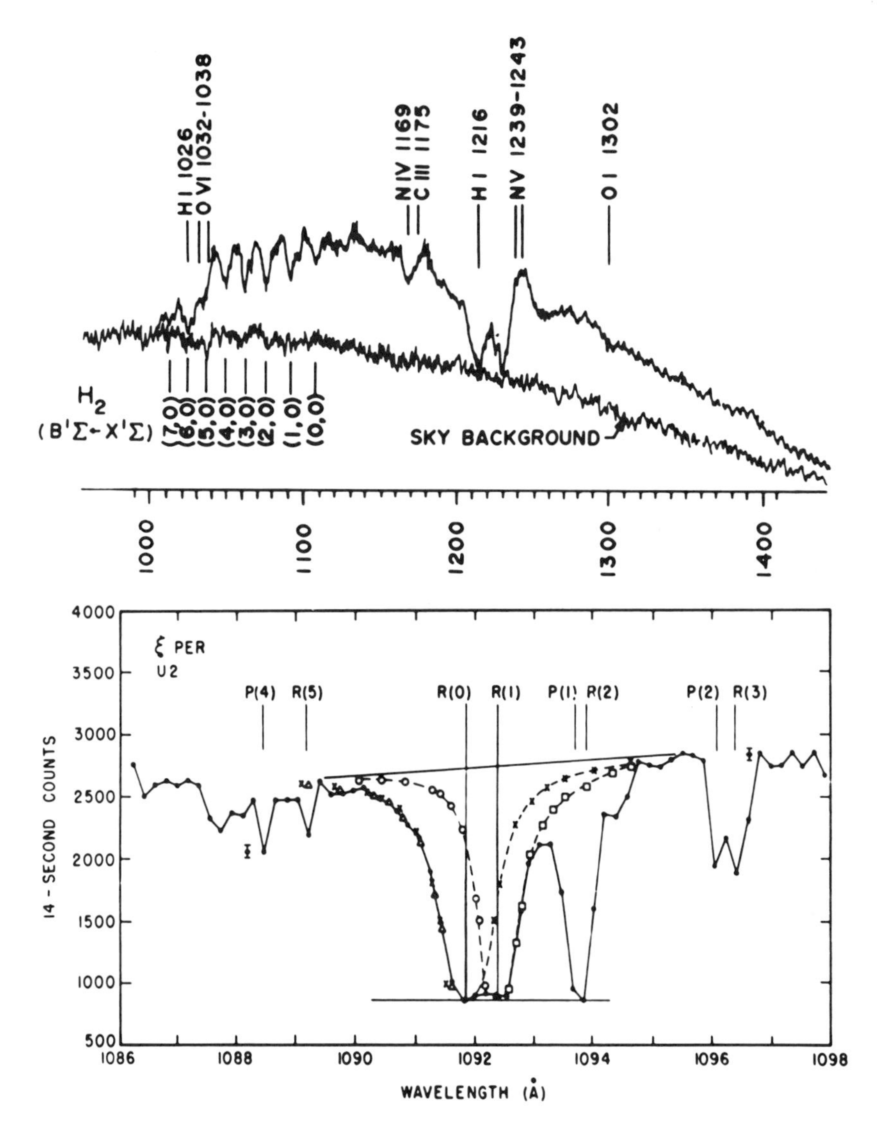

Figure 6.14

The spectrum shown at top marks the discovery of interstellar molecular hydrogen with the detection of its characteristic absorption between 1,000 and 1,100 Ångstroms. The lower figure represents a section of the H_2 spectrum of Xi Persei obtained, with higher spectral resolution, by the Copernicus satellite. (Courtesy of G. Carruthers, Naval Research Laboratory.)

Because the hydrogen atom itself provides such strong absorption in the interstellar medium, a major result has emerged from its observation: the great patchiness of the interstellar medium. When observing distant stars in the galactic plane, where one would expect an average hydrogen density to appear throughout, some directions are found to be almost empty of hydrogen. Indeed, the density may vary by as much as 1,000 times, even along neighboring lines of sight. This inhomogeneity is underscored by conditions in our "local" cosmic neighborhood, where the density of hydrogen is much lower than in the general interstellar medium. The sun is apparently located in a vast "interstellar hole." It is not clear how such density variations occur. Current theory suggests that the interstellar medium may be driven by the energy arising from supernova explosions. As the wavefront from such an explosion wends its way among the stars, secondary shocks form and the gas pushes and twists around the denser stars and clouds. Such a configuration might be compared to an intertwined nest of low-density tunnels; we may just happen to be in a local tunnel.

The second major result of ultraviolet spectroscopy has been the discovery of hot gases as well as cool material between the stars. The detection of interstellar absorption lines of five-times-ionized oxygen revealed that gas at temperatures of 200,000°K is prevalent. Since these lines are observed in the direction of hot stars, it was possible that the hot gas surrounded a star and was not a part of the general interstellar medium. Hot stars have fast, massive winds, as noted earlier, and these winds must run into the cooler interstellar medium, causing a shock that is somewhat akin to an interstellar bubble. However, additional surveys showed that the velocities of these oxygen lines did not correlate with stellar-wind velocities, as would be expected if a causal relation existed between the star and the high-temperature lines. Moreover, the oxygen velocities had a distribution that corresponded to a large-scale galactic distribution of a hot gas. Finally, the most recent x-ray images of stars also failed to show any extended circumstellar emitting regions that would result from an interstellar bubble.

An extended distribution of hot gas exists even above the plane of our galaxy, as is dramatically illustrated by the ultraviolet spectrum of a star in another galaxy, the nearby Large Magellanic Cloud. In this case, the astronomer has the advantage that the Large Magellanic Cloud is moving away from our galaxy at about 150 kilometers per second. This motion serves to separate features arising in the two galaxies. A double set of absorption features is found at widely separated velocities (figure 6.15). The low-velocity features (less than about 140

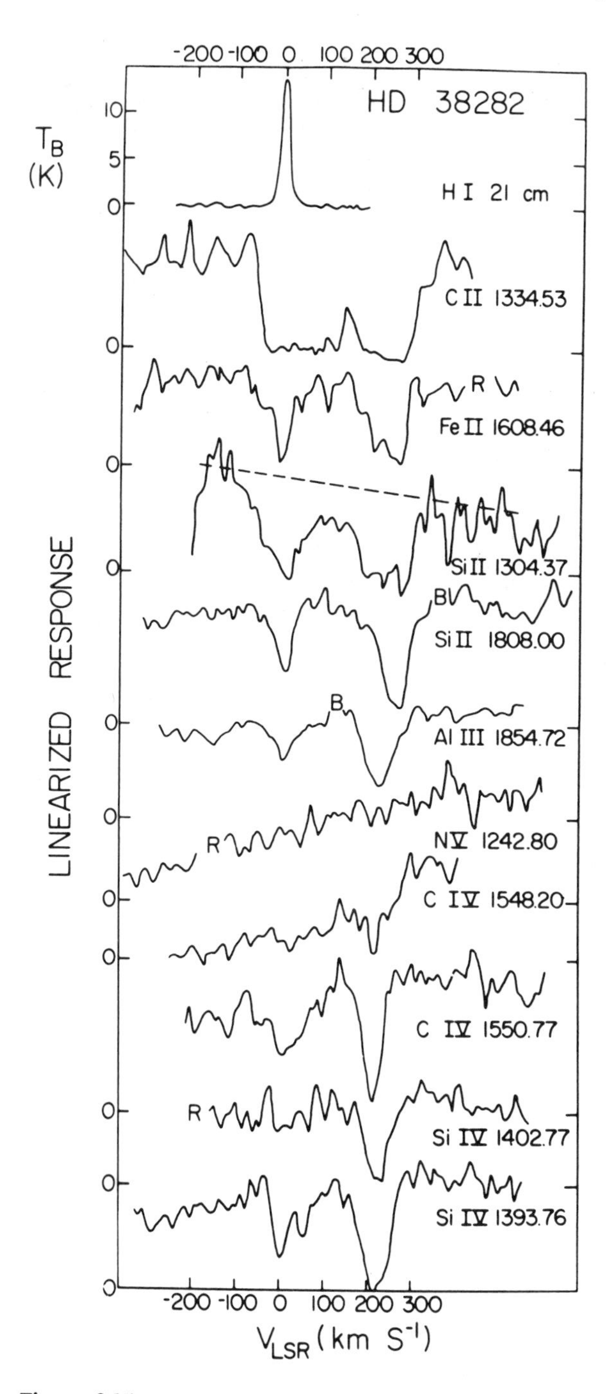

Figure 6.15
Ultraviolet absorption spectra toward the direction of a star in the Large Magellanic Clouds. Neutral hydrogen in our galaxy yields the emission profile shown at the top. Note the double set of components in the other ionized molecules, representing the double absorption. (Courtesy of B. D. Savage)

Figure 6.16
The Andromeda galaxy in optical light. This is a spiral galaxy thought to be similar to our own. (Harvard College Observatory photograph)

kilometers per second) arise from hot gas in our own galaxy; the high-velocity absorption results from material in the Large Magellanic Cloud. Analysis of the strength and velocity distributions of the lines in several directions shows that our galaxy has a halo of hot gas. This halo extends 26,000 light years—a distance comparable to that from the sun to the galactic center, but in a direction perpendicular to the plane of the spiral arms that form the disk of the galaxy. Since the halo of hot gas has been observed only in those directions, toward the few distant stars bright enough to be detected by the IUE satellite, it is not possible to say if the halo surrounds the entire Milky Way. In fact, there are some indications that the halo effect may be a special characteristic of that material lying in the direction of the Large Magellanic Cloud.

What makes a halo of hot gas so far above the plane of a galaxy? Several astronomers have suggested that it may bubble up from the plane. Since many galactic nuclei show evidence of mass ejection, perhaps a fountain of gas is expelled from the galactic center and then rains down over the disk. Or, perhaps forces in the spiral arms of the disk heat and propel material up and out of the plane. It will be important to determine the extent of our halo in all directions to understand the sources of this gas as well as to interpret the presence of absorption features in the spectra of other galaxies and, in particular, quasars.

Galaxies

Many individual stars in our own galaxy and in nearby galaxies, such as the Magellanic Clouds, are accessible with ultraviolet spectroscopic techniques, particularly those strong enough to resolve the spectral lines of specific elements. The more distant galaxies are accessible only with very long exposure times (around 16 hours) and even then only with the lowest resolution of the IUE satellite. Even so, some intriguing results have emerged. For example, the distribution of spectral energy in the continuum radiation can give clues to the temperatures of the stars present in a galaxy. If the temperatures are high, there must be a young, hot stellar population. If the temperatures are low, there is an evolved population of cool stars.

Ultraviolet spectroscopy often presents a surprise. The optical photograph of the galaxy Markarian 36 (figure 6.17) shows a neat little oval blob. There are no signs of any spiral arms of stars such as are found in our galaxy or in Andromeda. Optically, Markarian 36 appears to be an old galaxy in which all the stars have evolved. But the ultraviolet spectrum shows something quite different. The spectrum reveals much high-energy emission, indicating the presence of young, hot stars. Moreover, the absorption features are similar to those found in young, hot supergiant stars in our own galaxy. A theoretical calculation of the expected ultraviolet emission shows an excess of high-energy radiation. To fit this observed radiation with a theoretical model, we must assume that Markarian 36 has formed new stars within the last 10 million years. Observations like this suggest that star formation is an ongoing process in the universe.

Another landmark ultraviolet observation was the ultraviolet spectroscopy of the so-called double quasar. Ground-based radio and optical observations identified what appeared to be two practically identical quasars separated by only 6 seconds of arc in the sky. More astounding was the subsequent discovery that the optical spectra of the two objects were also identical. "Twin quasars" side by side in the same section of the sky seemed an unlikely chance occurrence. It was suggested that the two objects were really the same object, and the double image was the result of gravitational splitting of the quasar's light by an intervening massive object acting as a lens. A deep search with intensified optical cameras revealed the presence of a large, faint galaxy

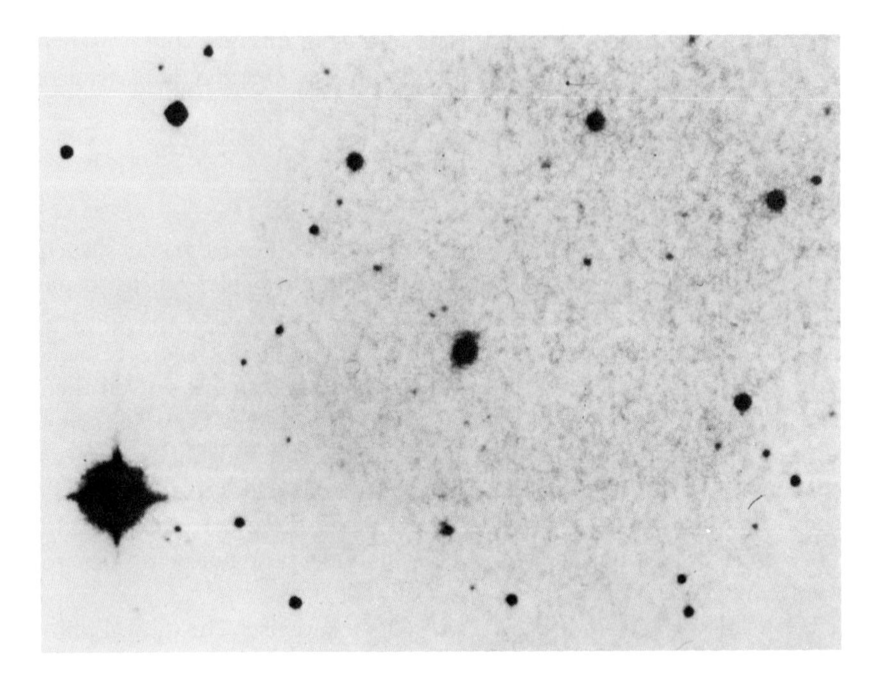

Figure 6.17
Optical photograph of the region containing the galaxy Markarian 36. The galaxy is the oval blob in the center of the field. (Courtesy of J. Huchra)

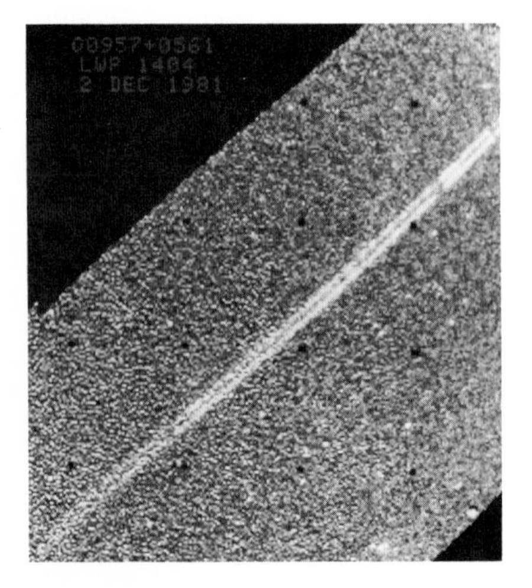

Figure 6.18
The ultraviolet spectrum of the double quasar (Q0957+561) as recorded by IUE at wavelengths from 2,200 to 3,000 Ångstroms. The spectra of both A and B components show Lyman α emission and absorption, which are red-shifted to about 2,900 Ångstroms. (NASA/Center for Astrophysics photograph)

lying between the quasar and the Earth to create a "gravitational lens" that was responsible in part for the double image.

One of the predictions of the gravitational-lens hypothesis for the double quasar is that the ratio of the energy in the two quasars should be the same at all wavelengths. In December 1979, measurements at radio, optical, and ultraviolet frequencies showed that the spectral components of the two images had the same ratio, thus providing more evidence for a gravitational-lens effect. Ultraviolet spectra obtained with IUE (figure 6.18) clearly show the red-shifted Lyman-α emission of the components.

At first it was thought that the radiation from the two images was constant; however, accumulating measurements at radio, optical, and ultraviolet frequencies suggest that some variation is present, and the variation seems to be greatest in the ultraviolet region. Additional information concerning the configuration of the intervening massive object can come from studying these variations over time. In particular, a measurement of any flux variation between the components can reflect the difference in travel time of light along the two lines of sight. If the images vary differently in one energy band than in the others, it could indicate that the quasar image is being lensed by low-mass stars in the halo of the intervening galaxy. Such observations provide a direct probe of fundamental physical phenomena on a cosmic scale.

The future of ultraviolet astronomy

The extent of this chapter demonstrates the influence of ultraviolet astronomy on all areas of modern astrophysics. Yet the space instruments available to date have allowed only the strongest ultraviolet sources to be observed. Major advances in space instrumentation are planned for the near future. In the mid-1980s NASA should launch the Space Telescope, a 2.4-meter telescope facility that represents the first permanent optical and ultraviolet space observatory. It will contain both imaging and spectroscopic instruments capable of observations at wavelengths down to 1,200 Ångstroms. These instruments can be removed and replaced with others by astronauts in space to provide at least two decades of scientific operation. A national Space Telescope Science Institute has been created to oversee the use of the Space Telescope and to act as a focal point for its scientific research.

Smaller satellites are to be dedicated to specialized tasks. The Extreme Ultraviolet Explorer, also to be launched in the mid-1980s, will survey the unexplored wavelength region from about 100 to 900 Ångstroms, in which nearby hot stars and the coronas of cool stars radiate profusely. Another satellite under consideration would perform very-high-resolution spectroscopic observations of distant objects in the wavelength region 900–1,200 Ångstroms, where interstellar gas, intergalactic gas, and galactic halos can be detected.

Such missions are exciting prospects. They will enable exploration of distant galaxies with the spectroscopic tools necessary to achieve a physical understanding of the components of the universe.

Reading

Bahcall, John, and Lyman Spitzer. "The Space Telescope." *Scientific American* 247 (1982): 40.

Beyond the Atmosphere: Early Years of Space Science. NASA report SP-4211. Washington, D.C.: Government Printing Office, 1980.

de Boer, K. S., and B. D. Savage. "The Coronas of Galaxies." *Scientific American* 247 (1982): 54.

Hanle, P. A., and V. Del Chamberlain, eds. *Space Science Comes of Age: Perspectives in the History of the Space Sciences.* Washington, D.C.: Smithsonian Institution Press, 1981.

Kondo, Y., J. M. Mead, and Robert D. Chapman, eds. Advances in Ultraviolet Astronomy: Four Years of IUE Research. Proceedings of a symposium held at NASA Goddard Space Flight Center, Greenbelt, Maryland, March 30–April 1, 1982. NASA conference publication 2238.

NASA. A Meeting with the Universe: Science Discoveries from the Space Program. Report EP-177. Washington, D.C.: Government Printing Office, 1981.

Spitzer, Lyman, Jr. *Searching Between the Stars.* New Haven: Yale University Press, 1982.

7

An X-Ray Portrait of Our Galaxy

Jonathan E. Grindlay

One of the most dramatic windows on the universe to be opened by space astronomy is the x-ray band of the spectrum. In fact, x-ray astronomy—the detection and study of cosmic sources of x rays—was among the first kinds of astronomy to be done from space. The electromagnetic spectrum, of which visible light is but a tiny slice, extends from the radio waves (the farthest to the red side of visible light) to the x rays and the gamma rays (the farthest to the blue side). Only certain narrow bands of this vast range of natural radiation are transmitted through the Earth's atmosphere and ionosphere to the ground. Most of the spectrum is absorbed at varying depths in the atmosphere, and astronomical observations are possible only from high-altitude balloons, sounding rockets, or satellites. In particular, x rays from cosmic sources are absorbed in the atmosphere, and thus observations of the sky at x-ray wavelengths must be carried out from space. These observations, in their short history of 20 years, have revealed a remarkable portrait of our galaxy and beyond.

The basics of x-ray astronomy

X rays are more energetic than visible light, just as blue light is more energetic than red light. As the energy of an electromagnetic wave increases, its wavelength decreases in exact proportion and the light displays more the properties of a particle than of a wave. "Particles" of electromagnetic radiation are called photons, and each single x-ray photon arriving from a celestial object can be detected. A variety of detectors are now used in x-ray astronomy. In all of these detectors, the x ray interacts with matter (either a gas or, more recently, a semiconductor solid composed primarily of silicon) and its energy is liberated as electric charges, which are collected and amplified to produce an electrical pulse. The number of these electrical pulses counted per unit time measures the x-ray brightness of the object under study; the distribution of pulse amplitudes, which are proportional to x-ray energy, measures its spectrum.

Both brightness and spectrum are fundamental quantities for describing the characteristics of a cosmic x-ray source. The third fundamental property is its position in the sky. If the x-ray source is pointlike (as is a star), then its shape and its angular size are not measurable and the position alone must be used to identify the source with an object detected at other wavelengths, such as an optical counterpart. If the x-ray source is extended (as is the hot gas swept up in the remnant of a supernova explosion), then its shape and distribution of x-ray surface brightness are important to a physical understanding of the source. The location of an x-ray source in the sky, even if only approximate, is much easier to measure than its brightness distribution, or x-ray picture, which must be concentrated and imaged on a position-sensitive x-ray detector. This became possible, for nonsolar x-ray photographs, with the launch of the Einstein (HEAO-2) Observatory on November 13, 1978. Before describing this marvelous instrument, which produced more than 7,000 x-ray images in its 2½-year lifetime, it is worth recalling the early history of x-ray astronomy, which provided the stimulation and motivation to build and operate the Einstein Observatory. The earlier x-ray-astronomy satellites also provided the first glimpses of the most exotic and energetic objects in our galaxy, neutron stars and black holes.

The early discoveries

X-ray astronomy has a short but exciting history. The first detection of a cosmic (i.e., other than the sun) source of x rays was made accidentally in 1962 by a sounding rocket launched to search for lunar x rays produced by the bombardment of energetic particles from the sun. The source was located in the constellation of Scorpio and, by the naming convention of radio astronomers, was called Sco X-1. This same first rocket flight, which exposed the Geiger-counter-type x-ray detector to space for only 5 minutes, also discovered a diffuse background of cosmic x rays. This is now known to be remarkably uniform and constant in all directions of the sky outside our galaxy. Its origin is still the subject of intense interest and controversy in high-energy astronomy. That both the diffuse background and a cosmic point source were discovered on the first x-ray-detecting rocket flight is a testament to the skill and scientific insight of the pioneering experimenters, led by Riccardo Giacconi and Bruno Rossi.

Over the next 8 years, many rocket experiments, with increasingly larger and sophisticated x-ray detectors, were carried out by a number of groups. More than 30 bright (point) sources of x rays in our galaxy, and several of the brightest sources outside our galaxy, were discovered. Most sources were located only to a precision of about 0.5°, and optical identifications were not possible.

In December 1970 the first satellite devoted to x-ray astronomy was launched. This satellite was officially named SAS-1 (Small Astronomy Satellite 1), but the scientists responsible for the experiment (again headed by Giacconi) renamed it Uhuru, which means "freedom" in Swahili, in honor of Kenya's Independence Day. (The satellite was launched on that day from a platform in the Indian Ocean off the Kenya coast in order to achieve an orbit about the equator, where the background "noise" in the x-ray detectors would be minimal.) Uhuru surveyed the entire sky for the first time and cataloged several hundred sources of x rays. A map of the distribution on the sky of these x-ray sources is shown as figure 7.1. The size of each dot corresponds to the apparent intensity of the x-ray source in the energy band from 2 to 6 keV. (One keV, or kilo-electron volt, is about 500 times the energy of the visible-light photons we see, which have energies of about 2 electron volts.)

The "flattened globe" grid lines on the map in figure 7.1 represent the galactic coordinate system used by astronomers to denote the positions of objects in the galaxy. The horizontal midline represents the plane of the Milky Way, and the curved horizontal lines above and below are lines of constant latitude (30° and 60°) as seen from Earth's perspective in the solar system. The curved vertical grid lines denote galactic longitude, with 0° (the center of the map) corresponding to the center of the galaxy and 180° (the extreme left and right sides of the map) corresponding to the direction exactly opposite to the galactic center. Most of the brightest x-ray sources are clustered along the plane of the galaxy and are grouped around its center. As we learned with the Einstein satellite, the spatial distribution of bright x-ray sources in our galaxy looks very much like that of the bright sources in the Andromeda galaxy, the great spiral galaxy closest to us. This galaxy, with the positions of its x-ray sources (also measured with the Einstein Observatory) superposed, is shown in figure 7.2.

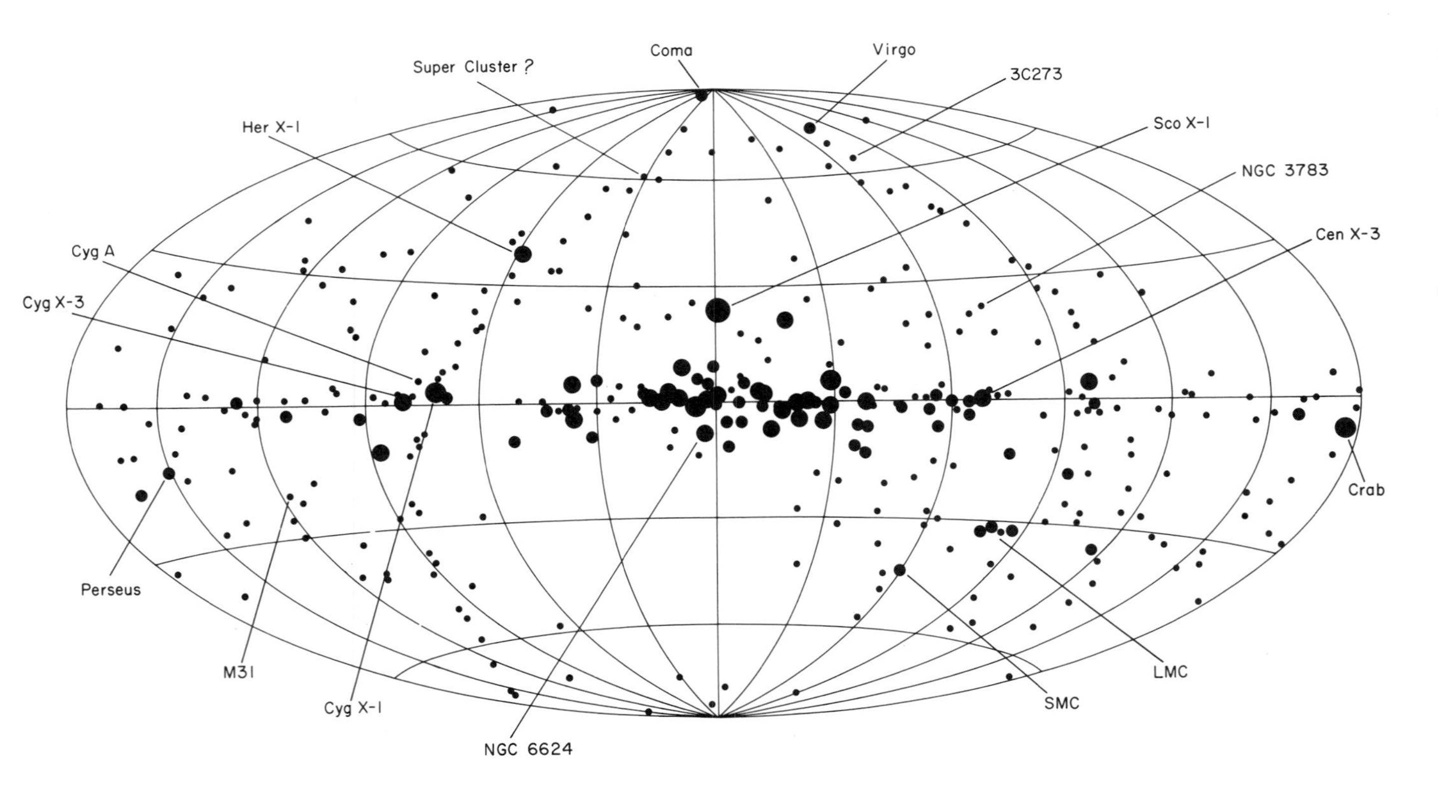

Figure 7.1
Map of the x-ray sources detected by Uhuru, the first satellite dedicated to x-ray astronomy. The relative brightness of each source is indicated by the size of the dot; the grid lines are the standard galactic coordinate system used by astronomers. (Courtesy of Smithsonian Astrophysical Observatory)

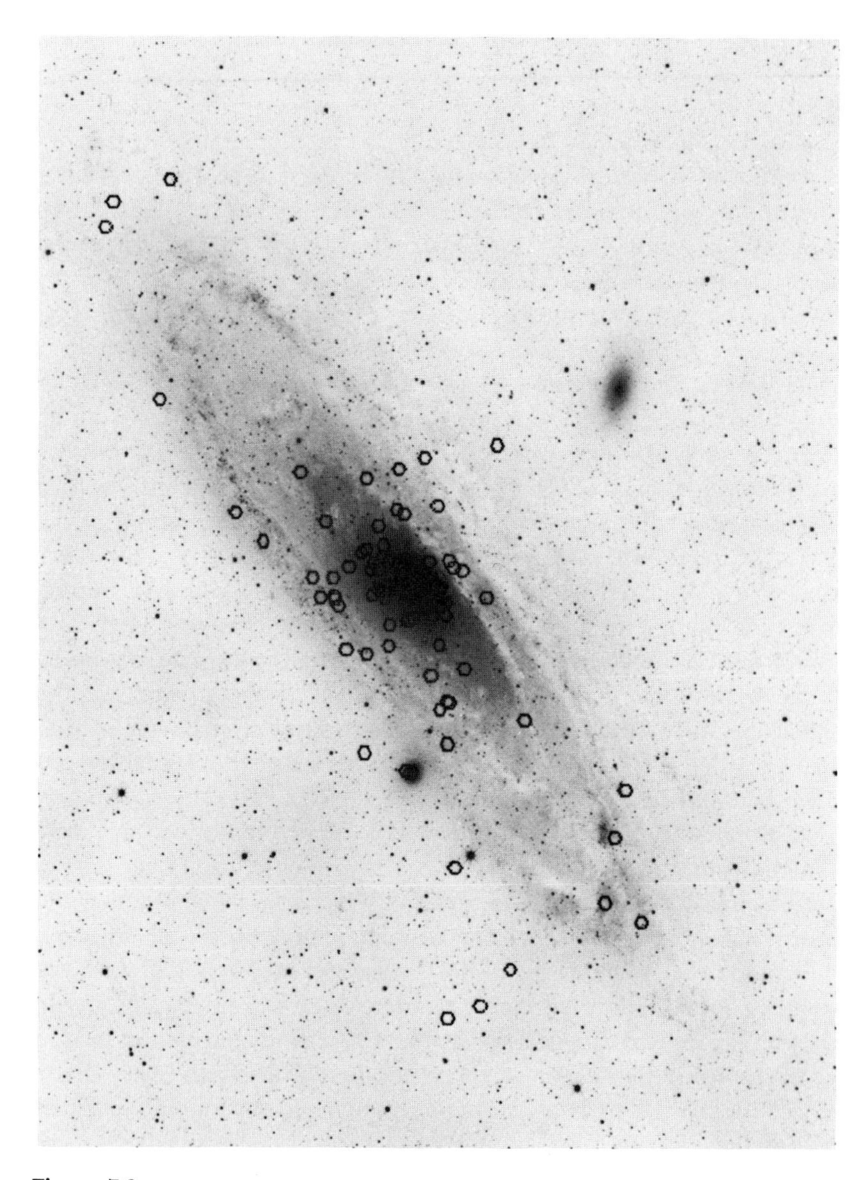

Figure 7.2
Distribution of x-ray sources detected by the Einstein Observatory in the giant spiral galaxy M31. This galaxy, the Andromeda nebula, is very similar in size and structure to our own. (Courtesy of Smithsonian Astrophysical Observatory)

Discovering x-ray binaries: Neutron stars and black holes

Detailed observations of the bright x-ray sources in our galaxy with Uhuru revealed that, in general, they are highly variable in their x-ray emission. From the outset variability seemed to be the rule, rather than the exception as astronomers had become accustomed to supposing from centuries of visible-light observations of stars. In several cases, Uhuru was able to detect highly regular variations in the x-ray flux from a particular source. For example, the source Centaurus X-3 was seen to pulse on and off every 4.84 seconds and to switch its pulsations on and off every 2.1 days (figure 7.3). The observed behavior suggested that Cen X-3 was a pulsar and a member of a binary system, with the x-ray source periodically eclipsed by the stellar companion about which it orbited. Striking proof that this x-ray source was in orbit about a companion star comes from the precise timing of the arrival of the individual 4.8-second pulses: The pulse arrival times are smoothly modulated with the 2.1-day period. This can only be due to the changing Doppler shift (analogous to the changing pitch of a train whistle as the train passes by) of the pulsar "clock" as it periodically moves toward and then away from Earth. Such periodic motion is the signature of a member of a binary star system. The binary picture also provided a framework on which to construct theoretical models for other similar x-ray sources.

The regular pulsations of Cen X-3 also implied that the x rays were somehow produced near the surface of an object spinning with a period equal to the pulsation period. Only rotation, not vibration, could provide the precise stability of the pulsation periods observed. However, the rapid rotation rates (i.e., short pulse periods) required that the spinning x-ray source be highly condensed. An ordinary star would fly apart from centrifugal force at the 4.8-second spin period of Cen X-3. The x-ray pulsars must, therefore, be collapsed and very dense objects—either white dwarfs or neutron stars, two possible end products of stellar evolution.

When a dying star expels its lighter outer atmosphere, its core of heavier materials, now no longer supported against gravity by internal energy sources, collapses to become an extremely compact object with a "solid" surface. The collapse can be halted at a radius comparable to that of the Earth if the core mass is less than 1.4 solar masses; the resulting object is known as a white dwarf. But a collapsing stellar core with mass between 1.4 and about 3 solar masses may fall in on itself in a few seconds to a radius of only about 10 kilometers. These are the neutron stars, with the material (almost entirely neutrons, since

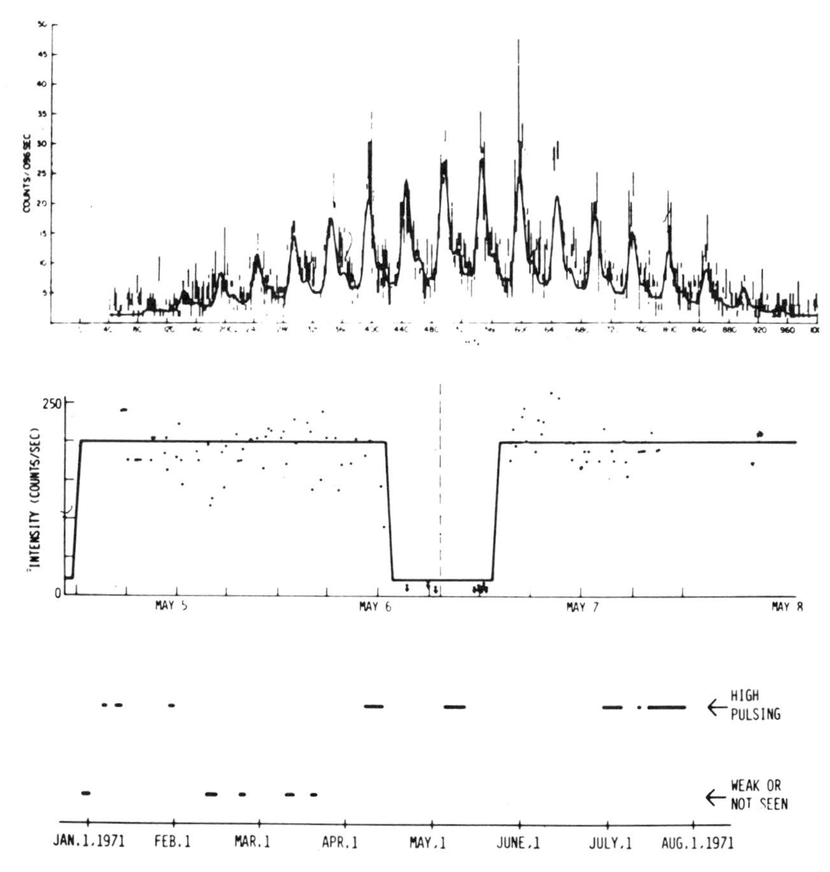

Figure 7.3
(Top) X-ray pulsations at 4.8-second period. (Middle) Binary eclipses at 2.08-day period. (Bottom) Long-term intensity variations of the binary x-ray pulsar Centaurus X-3. (Courtesy of Smithsonian Astrophysical Observatory)

the original protons and electrons are squeezed together) of the entire star compressed to the fantastic density of nuclear matter, or about 300 million tons per cubic inch. However, the collapse of very massive stars (those whose stellar cores have mass greater than 3 solar masses) cannot be halted even by nuclear forces in the neutron-star stage. Instead, gravity dominates all, and the objects literally collapse out of sight to become black holes—objects so dense that not even light can escape their "surface." In fact, a black hole's "surface," which would have a radius of about 10 kilometers for a 10-solar-mass black hole (a black hole's radius is directly proportional to its mass), is really a boundary and not a solid surface, so that nothing can be tied to it. The surface of a neutron star or a white dwarf, by contrast, can anchor a magnetic field, which need not be aligned with the rotation axis of the star. Thus, the rotation of a neutron star or white dwarf (but not a black hole) could produce x-ray pulses if the x rays were somehow produced near the "north" and "south" poles of the rotating star and the magnetic and rotation axes were misaligned. In the case of Cen X-3 and the other bright x-ray pulsars discovered by Uhuru, additional observations have made it clear that the rotating, magnetized, compact objects giving rise to the x-ray pulses are neutron stars and not white dwarfs.

But how are the x-ray pulses produced near the magnetic poles of a rotating neutron star? The answer is still uncertain in its details, but the broad picture is clear. Matter is gravitationally captured, or accreted, by the neutron star from the atmosphere of the normal stellar companion in the binary system. In the case of Cen X-3, the normal stellar companion was identified optically as an early-type massive star in which mass loss in the form of a wind is occurring. Some of the matter in this wind is captured by the intense gravity of the neutron star and falls to its surface along the lines of the magnetic field. These lines converge at the magnetic poles; hence, the matter tends to accrete there. As the matter (mostly hydrogen and helium) crashes down on the surface of the star, it must give up the energy of its motion into radiation. This radiation is in the "hard" x-ray or gamma-ray energy ranges, but it is converted, through repeated scatterings, into other, lower-energy radiation in the x-ray band. Again, the details of the radiation production are not well known, but the total energetics of the accretion process are consistent with the observed x-ray luminosities. The x-ray luminosities of binary systems are typically from 10^{36} to 10^{37} ergs per second—more than a thousand times the power that our sun radiates in the entire electromagnetic spectrum.

Whereas the x-ray pulsars found with Uhuru and with x-ray detectors on followup satellites could be determined unambiguously to be members of binary systems, many others of the bright sources mapped in figure 7.1 were much more enigmatic. One of these sources, Cygnus X-1, was found to be particularly puzzling. Unlike virtually all the others, it undergoes sporadic variations in x-ray intensity on all time scales down to a few milliseconds. Clearly accretion was once again involved to produce the x rays, but the peculiar flickering emission suggested matter falling onto a black hole rather than a neutron star. The binary nature of Cyg X-1 was finally revealed when an optical counterpart was discovered fortuitously after an astrophysically interesting change in Cyg X-1 occurred.

In March 1971, Cyg X-1's average x-ray intensity decreased abruptly and a relatively bright radio source appeared. The radio source could be positioned with arc-second accuracy (unlike the x-ray source, which Uhuru could locate only to a fraction of a degree), and appeared coincident with a bright early-type star. Doppler shifts in the stellar spectrum revealed that indeed the star was a member of a binary system with a 5.6-day period. The mass of this star, the binary companion and gas supplier to Cyg X-1, could be estimated from its spectrum to be about 25 solar masses. More important, the magnitude of the periodic Doppler shifts in the stellar spectrum and the range of likely orientations of the binary system made it possible to "weigh" the x-ray source, too. From its motions, the unseen orbiting companion was estimated to have a mass of at least 9 solar masses. This is much greater than the allowed maximum mass for a neutron star, and it provided the best evidence to date for the discovery of a black hole. (Cygnus X-1 would also be the first cosmic x-ray source imaged with arc-second resolution, by the Einstein Observatory in November 1978).

The Einstein Observatory: The first x-ray images

Before the Einstein Observatory our x-ray view of the galaxy was dominated by the bright accretion sources, both pulsars and black hole candidates. Several supernova remnants (the gaseous debris of exploded stars) were also known to be x-ray emitters, but their emission could not be mapped in much detail. A handful of truly stellar x-ray sources were also known, but these were all unusual (e.g., flaring) stars, and stellar x-ray emission was thought to be the exception rather than the rule. The x-ray spectra of all these types of galactic x-ray sources were poorly known, with mere hints of line emission for some sources and apparently smooth continuum spectra for most others.

X-ray emission was thought to be confined generally to the exotic objects and not to be a diagnostic of a wide range of astrophysical phenomena.

The Einstein Observatory changed all this, as its thousandfold increase in sensitivity revealed countless new x-ray sources as well as the fine details of previously known galactic sources. Officially called HEAO-2 (High Energy Astronomy Observatory 2) by NASA, the satellite was named the Einstein Observatory by the astronomical community in honor of the Einstein centenary. The x-ray telescope carried by the Einstein Observatory is shown in schematic form in figure 7.4. X rays were imaged by reflecting (at small grazing-incidence angles of about 1°) off two highly polished cylindrical mirrors, the first with a parabolic curvature and the second with a hyperbolic curvature. These two reflections brought the beam of incoming x rays to a common focus some 3.4 meters behind the mirror, where a "lazy Susan" assembly could rotate any one of four detectors into the focused x-ray image. Since the effective aperture of such an x-ray telescope is small (because of the glancing angles at which x rays can be reflected), the Einstein telescope contained four concentric mirrors. The outer mirror had a diameter of 0.6 meter.

The actual detectors at the telescope focus included two imaging "cameras" (position-sensitive detectors) and two spectrometers. The imaging detectors, which had either high (approximately 2 arc-seconds) or low (approximately 1 arc-minute) angular resoultion, were the high-resolution imager (HRI) and imaging proportional counter (IPC) detectors, respectively. The IPC had a greater intrinsic sensitivity over a broader energy band (0.2–4 keV) than the HRI, which had no energy resolution despite its much higher angular resolution. The spectrometers were similarly complementary: a low-spectral-resolution but high-sensitivity solid-state spectrometer (SSS) and a high-resolution but lower-sensitivity focal-plane crystal spectrometer (FPCS). The HRI and IPC detectors were provided by the Smithsonian Astrophysical Observatory (SAO), whereas the SSS was provided by the Goddard Space Flight Center (GSFC) and the FPCS by the Massachusetts Institute of Technology (MIT), all under contract to NASA. The Einstein Observatory was operated for its 21/2-year lifetime by a consortium of the x-ray groups at SAO, MIT, GSFC, and the Columbia Astrophysics Laboratory (CAL). The satellite was controlled by NASA from the GSFC.

Approximately a week after its launch on November 13, 1978, from the Kennedy Space Center by an Atlas-Centaur rocket, all the instruments and systems on the Einstein Observatory had been turned on and checked, and observations began. As already mentioned, the first

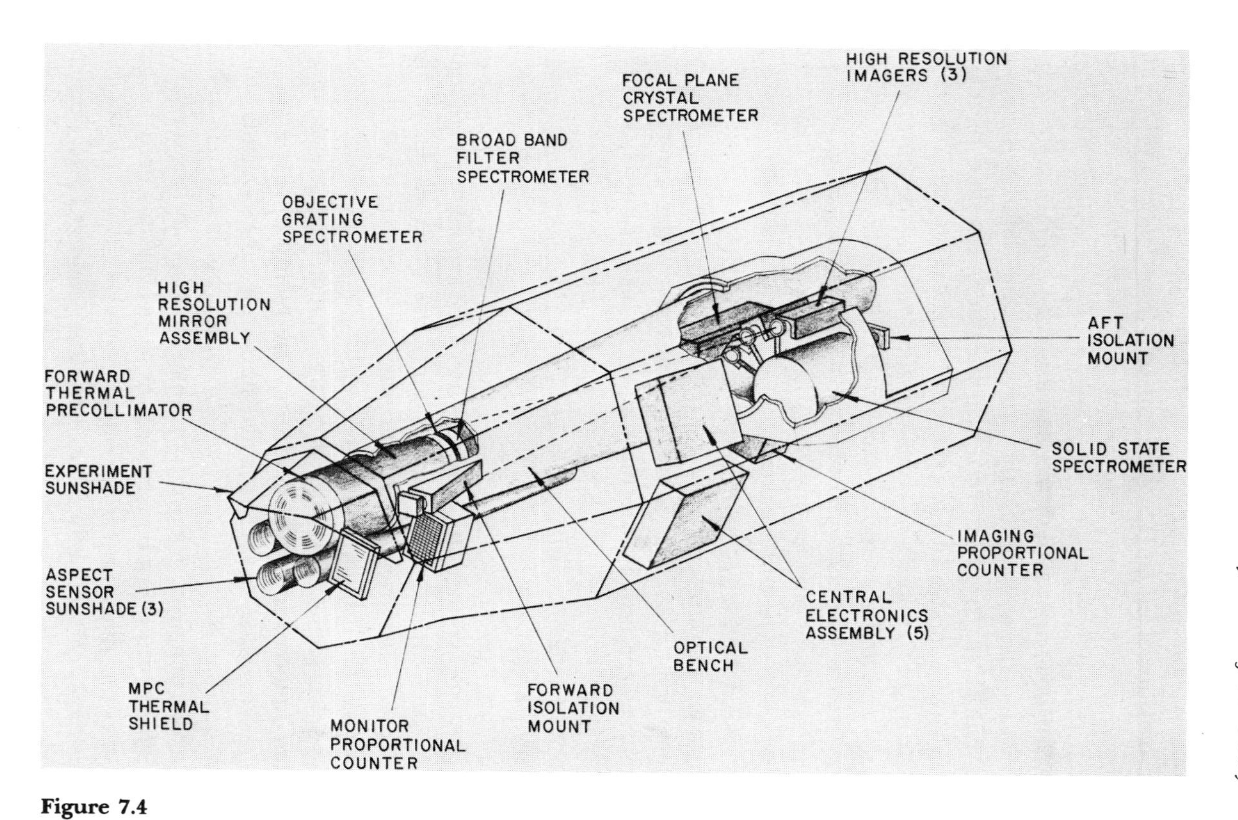

Figure 7.4
Schematic diagram of the Einstein Observatory. (Courtesy of Smithsonian Astrophysical Observatory)

actual observation was of Cyg X-1. The image, obtained with the HRI, showed an unresolved (at the approximately 2-arc-second resolution limit of the telescope) point source of x rays, with no other (weaker) source in the approximately 25-arc-minute field of view. Although this first image contained no surprises, it was an exciting moment for cosmic x-ray astronomy. Valuable new scientific data on the rapid time variations of Cyg X-1 were also obtained. The first of many surprises to be found with Einstein was evident shortly thereafter in an IPC image of Cygnus X-3, a binary system with a very short period (4.8 hours) and an x-ray source well studied by earlier detectors. As the x-ray image was displayed on the video display system at SAO, it was immediately obvious (figure 7.5) that the field also contained five other sources, much weaker in intensity. These "serendipitous sources" were the first of many such "extra" objects to be observed with Einstein; thousands were discovered in subsequent x-ray images. For observations at low galactic latitudes (near the plane of our galaxy) these sources are usually "normal" stars; at high latitudes they are primarily quasars.

X rays from normal stars

Given the spatial configuration on the sky of the serendipitous sources near Cyg X-3, their optical identification was relatively easy. The optical counterparts are indicated in figure 7.5, where they are labeled A–E; they are the relatively bright stars within the circles, which denote the most probable location of each x-ray source. These stars had all been well studied previously, since they are members of a group of massive, young, hot, recently formed stars called the "Cygnus OB-2 Association." (In fact, star C is one of the most luminous stars in our entire galaxy; in figure 7.5 it does not appear intrinsically bright relative to the other x-ray stars because it is very heavily obscured by gas and dust in the galaxy along the line of sight.)

Later studies of many other early-type stars observed with Einstein have shown that their fractional luminosity in x-ray emission (at about 1 keV) is very nearly 10^{-7}. This result was unexpected because early-type stars are not expected to have high-temperature coronas and thus should not emit x rays. (The base of the solar corona, the sun's thin, hot outermost atmosphere, which is visible only during solar eclipses, had been detected in x-ray observations before Einstein and had been regarded as a prototype stellar x-ray source.) In fact, the x rays probably arise from large mass outflows in the form of stellar winds from these stars. As the wind plows into the surrounding in-

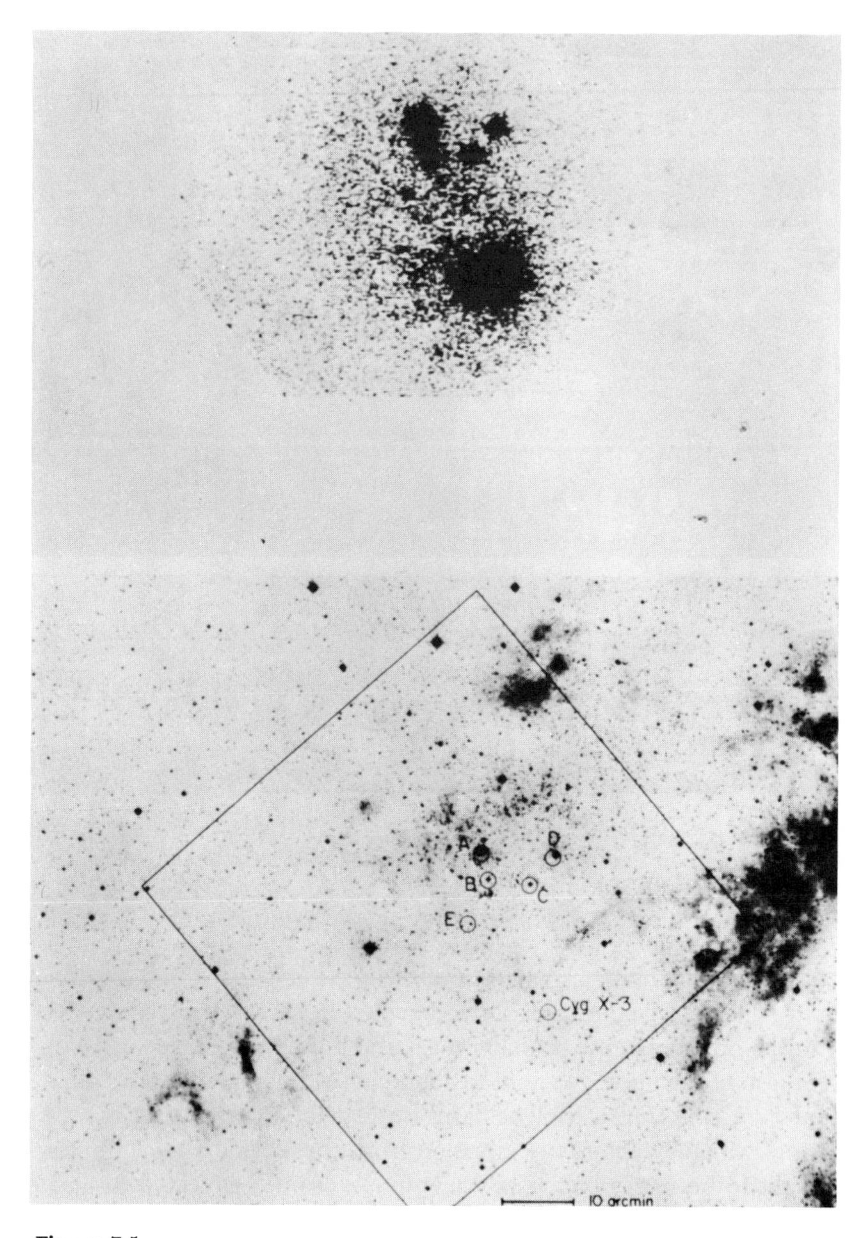

Figure 7.5
X-ray image (top) and optical image (bottom) of a region of sky (enclosed in the box) in the constellation Cygnus. In addition to the very bright x-ray source Cyg X-3, five much fainter sources (A–E) were discovered. The individual "speckles" in this and the subsequent x-ray images shown here are not individual "stars," but background events (x rays and particles) recorded by the detector. (Courtesy of Smithsonian Astrophysical Observatory)

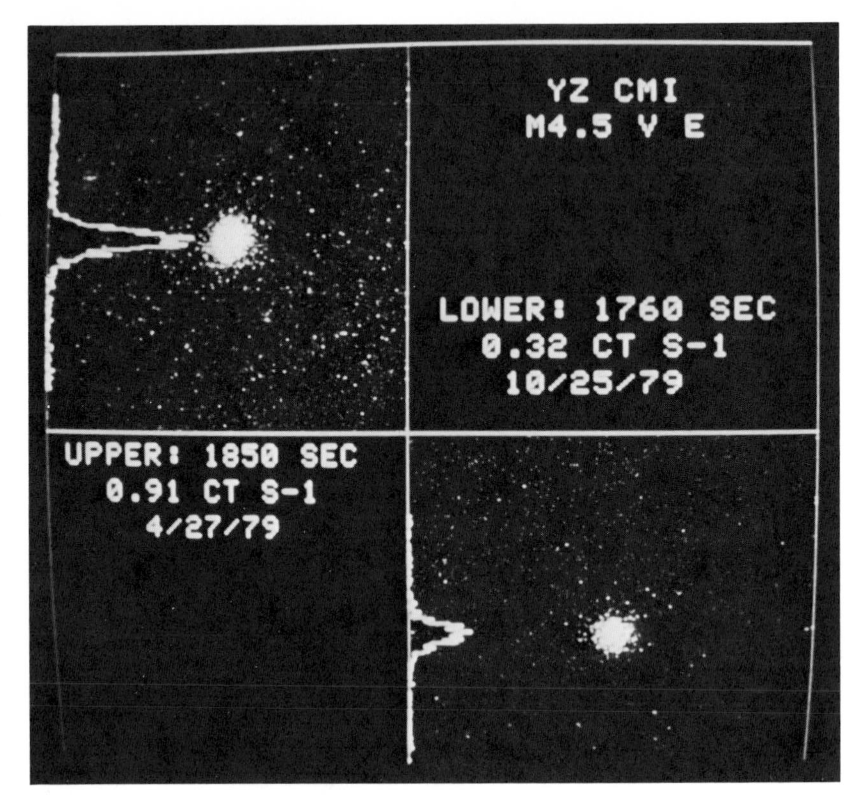

Figure 7.6
X-ray images of the flare star YZ-CMI, showing differing x-ray intensities due to flares.

terstellar medium, shock waves are set up, which heat the gas to the multimillion-degree temperatures detected in x-ray observations.

At the other extreme of stellar mass and luminosity, red dwarf stars were also not expected to be x-ray sources. Although these stars often produce x rays during violent flares (similar to large solar flares), the normal, or quiescent, x-ray emission from these faint, cool stars was thought to be negligible, since they, too, were not expected to have high-temperature coronas. But Einstein images of red dwarfs have shown that they are also prolific x-ray emitters. Almost 10 percent of the total energy radiated by these supposedly "dull" stars turns out to be in the x-ray band. One such x-ray source, the flare star YZ–Canis Minoris, which was one of the two detected prior to Einstein, is shown in figure 7.6. Quiescent x-ray emission with a luminosity of about 10^{28} ergs per second is shown in the lower half, while the upper half shows the increased emission during a flare lasting about 5 minutes.

Since these stars are the most numerous in the galaxy (most of the known stars closest to the sun, for example, are red dwarfs), their total contribution to the overall x-ray emission from the galaxy is probably very significant. Very likely, too, they contribute a large fraction of the diffuse soft x-ray background flux, which, unlike the diffuse flux at higher energies, must originate largely in our galaxy. Since red dwarfs have been postulated on other grounds to make up a major part of the "unseen" mass in the spherical halos surrounding galaxies, their integrated x-ray emission should be detectable in future, more sensitive x-ray observations of galaxies seen approximately edge-on.

The x-ray flux from red dwarf stars may originate in their highly convective atmospheres. Magnetic field loops may be formed in these bubbling atmospheres and may, through annihilation (or north-and-south-pole "reconnecting") liberate their energy into x rays. In addition, flares may arise from the reconnections of large-scale magnetic fields over huge "star spots" on the stellar surface. The discovery of quiescent x-ray emission from these stars has opened up a major field of astronomical study.

In certain regions of our galaxy, stars are being formed from great clouds of gas and dust. Since the most massive stars live only a few million years before exploding in supernova events (in which more energy—about 10^{51} ergs—is released in a few seconds than in the whole previous life of the star), stellar birth and death may be expected to occur in common regions of space. One such region is in the southern constellation of Carina, where the star Eta Carina was observed in the 1840s to brighten to become (for a few years) one of the brightest stars in the sky. The Einstein x-ray image (in iso-intensity contour form) of the Eta Carina complex is shown in figure 7.7, superimposed on an optical photograph of the same region. (The rectangle superposed on the image is a support structure on the detector.) Several bright point sources of x-ray emission are readily visible: Eta Carina itself (the lower left point of the four in the center), the bright open clusters of young stars Tr 14 and Tr 16, and several individual bright early-type stars. These x-ray stars are similar to those in the Cyg OB-2 association. In addition, there is large-scale diffuse emission (as indicated by the large-scale contours), which also shows the same band of absorption (by interstellar dust) apparent in the optical photograph as a curving band running below the complex of bright point sources. The diffuse x-ray emissions may arise from the overlapping shells of the remnants of earlier supernovas in this complex, and the combined effects of stellar winds from the many hot stars probably also contribute.

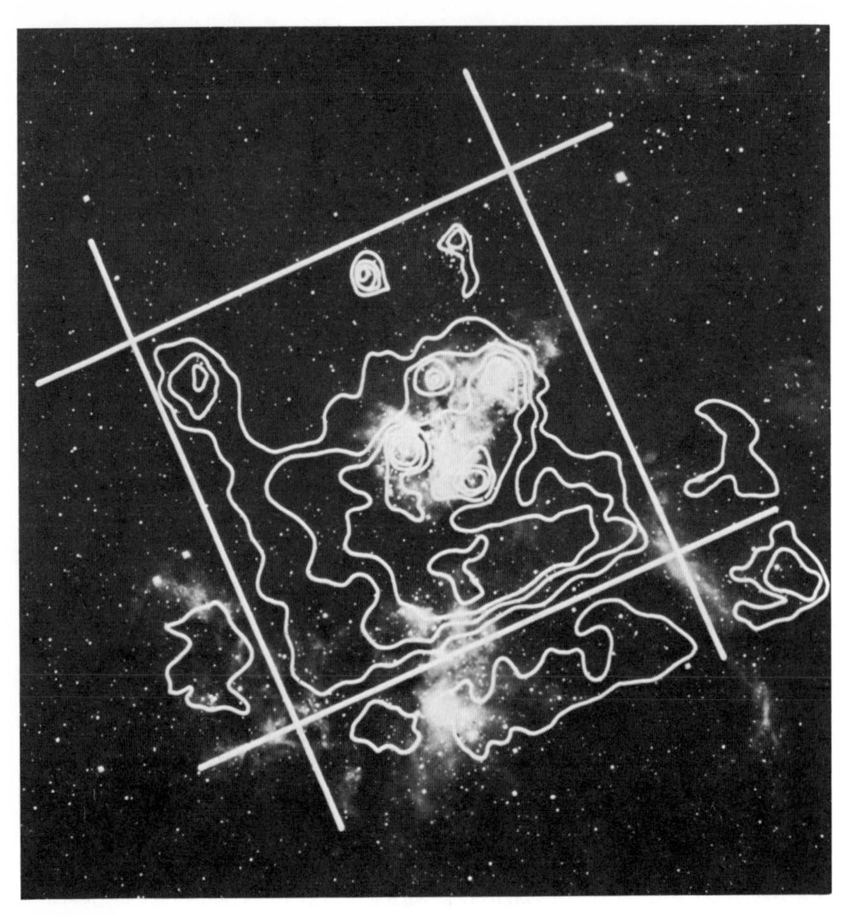

Figure 7.7
X-ray map of the Eta Carina nebula superimposed on an optical photograph of the region.

Pulsars and supernova remnants

The Crab nebula was the first galactic x-ray source to be identified optically, and it has been the subject of intense study at x-ray (and gamma-ray) energies ever since. The Crab is the remnant of a supernova explosion observed by Chinese astronomers in July 1054 A.D. and reportedly visible by daylight for nearly 3 weeks. The x-ray image of the Crab as seen by the Einstein Observatory is shown in figure 7.8. On approximately the same scale, the optical image of the Crab is also shown in figure 7.8. The x-ray emission is both diffuse and from a central point source: the 33-millisecond pulsar in the Crab nebula. The x-ray image of figure 7.8 shows the pulsar in its "on" and "off" phases by adding together all the x-ray photons recorded at each phase during successive 33-millisecond pulse periods. The optical image of the Crab is an integrated exposure and shows the pulsar at its time-averaged brightness. (The pulsar is the lower of the two faint stars near the center of the nebula.)

Spinning at 30 revolutions per second, the Crab pulsar is the third fastest known. It has been detected across the entire electromagnetic spectrum, from radio waves through the highest-energy gamma rays. Until the Einstein images, it was not known whether x-ray emission was also produced by the pulsar on a continuum background between the pulses. Figure 7.8 shows that, in fact, the pulsar disappears completely, or is indistinguishable from the surrounding nebula, during the pulsed "off" phase. (This sets important new limits on the rate at which neutron stars must cool after being born in a fiery supernova explosion, since any "continuous" emission from the pulsar itself would likely be due to thermal radiation from the very hot neutron-star surface.)

The diffuse x-ray emission in the Crab is due to synchrotron radiation, a process whereby electrons with very high energies spiral in a magnetic field and radiate x-ray photons as a result of their circular accelerations. As can be seen in figure 7.8, this radiation is not symmetric about the pulsar, although the pulsar is almost certainly the source of the energetic electrons. Moreover, the x-ray image of the Crab is much more asymmetric about the pulsar than is the optical image. This asymmetry is not understood, but it may reflect an anisotropy in the initial supernova explosion or in the surrounding medium. The resulting inhomogeneities in the magnetic field densities as well as the acceleration and propagation of the most energetic electrons in the nebula would then give rise to the asymmetric x-ray emission.

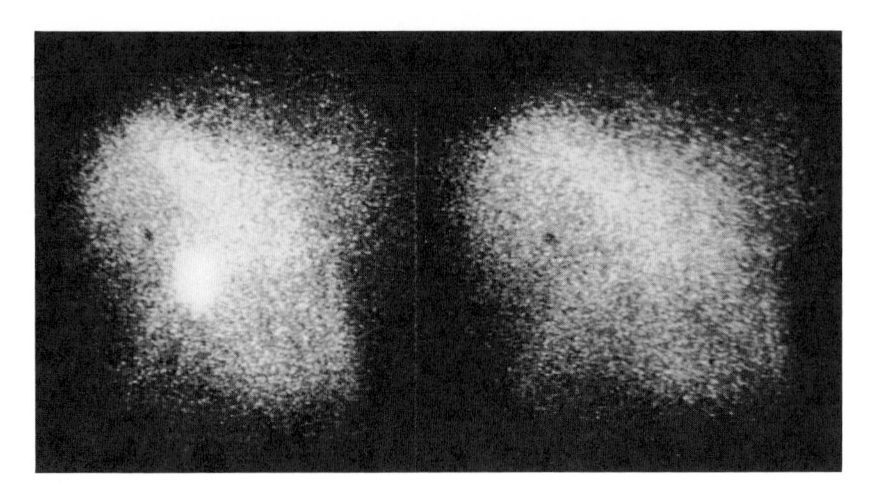

Figure 7.8
X-ray image of the Crab nebula, showing the pulsar in its "on" (left) and "off" phases.

Figure 7.9
Optical image of the Crab nebula.

Many other supernova remnants (SNRs) were imaged for the first time in x rays by the Einstein Observatory. These images are not only of great astronomical interest but are also among the most beautiful recorded by Einstein. Some, like the Crab, are filled shells with symmetry about only one axis. Others are much more round and shell-like, with empty centers. Still others are highly irregular and wispy, with structures apparently determined more by the clumpy interstellar medium than by the circular symmetry of the initial supernova blast. Only four SNRs (including the Crab) are known to contain x-ray pulsars, the neutron-star collapsed cores of the original stars. Two of these pulsars (in the Crab and the Vela nebulas) were originally discovered as radio pulsars in the late 1960s. Pulsars were then sought with radio telescopes in other SNRs to prove further the connection between neutron stars and supernovas, but none were found. Einstein added the other two pulsars now known in SNRs and thereby strengthened considerably the theory that neutron stars, as well as at least some x-ray binary systems, originate with supernovas.

The composition and the physical conditions (temperature and density) of the gas in SNRs also have been studied intensively with the Einstein Observatory. This has been made possible by the unique x-ray-spectroscopy capabilities of the two spectrometers on board Einstein. The spectra have revealed rich line emission from the hot, thin gas clouds heated by the shock waves set up in the expanding supernova shell. An example of such a spectrum is shown in figure 7.10, where x-ray emission from oxygen ions in the Puppis A SNR at a temperature of several million degrees is apparent. This highest-resolution, most complete spectrum of any cosmic x-ray source yet recorded was obtained with the FPCS (a Bragg crystal spectrometer) on the Einstein Observatory. The spectrum was obtained from only the brightest part of the very complex and beautiful diffuse emission evident in the montage of several overlapping HRI images shown in figure 7.11. The x-ray spectra of SNRs indicate that the previous assumption of equilibrium in the various stages of ionization of the hot gas in a SNR was wrong. The complex spatial structures apparent in figure 7.11 suggest that substantial spatial structure also exists in the ionization balance of the gas. Future high-spatial-resolution x-ray spectrometers are needed to attack these important questions.

Galactic-bulge x-ray sources

The very brightest galactic x-ray sources—those clustered near the central bulge region of our galaxy (figure 7.1)—continue to present

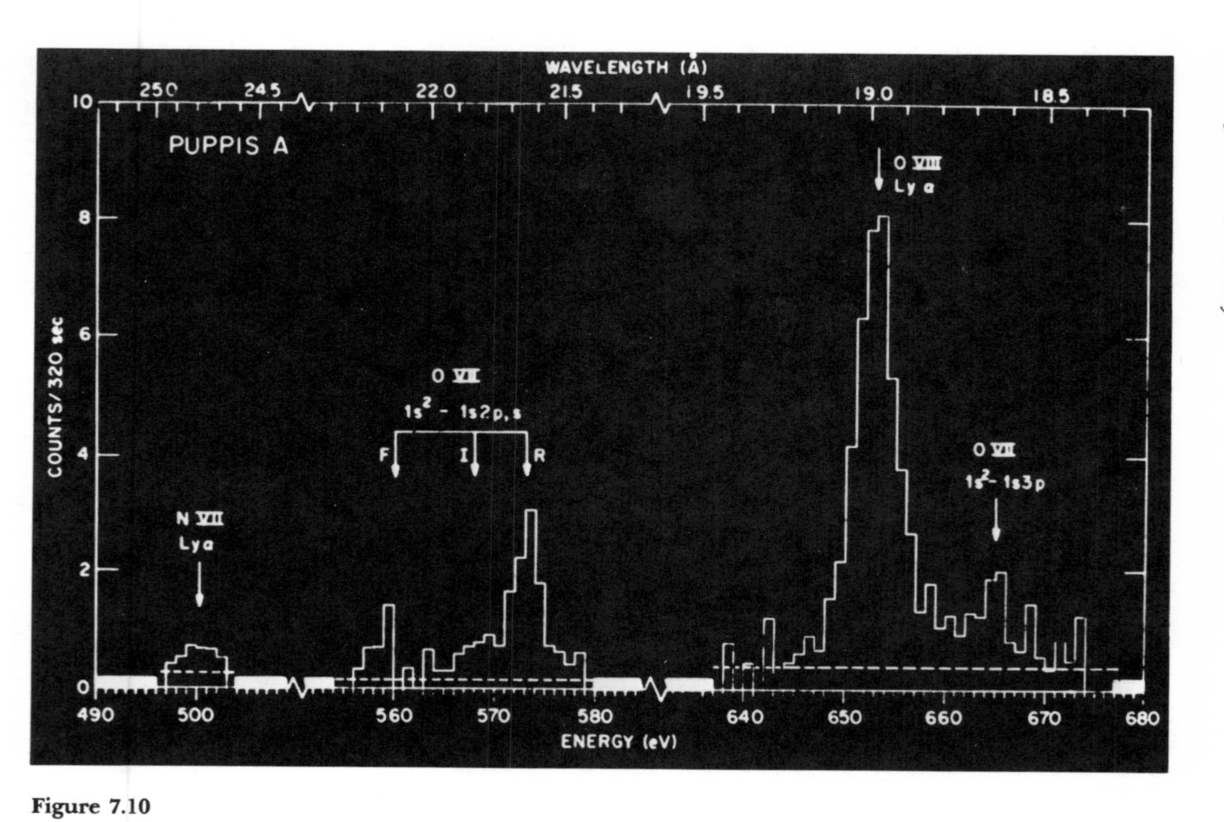

Figure 7.10
X-ray spectrum of the brightest knot in the supernova remnant Puppis A.

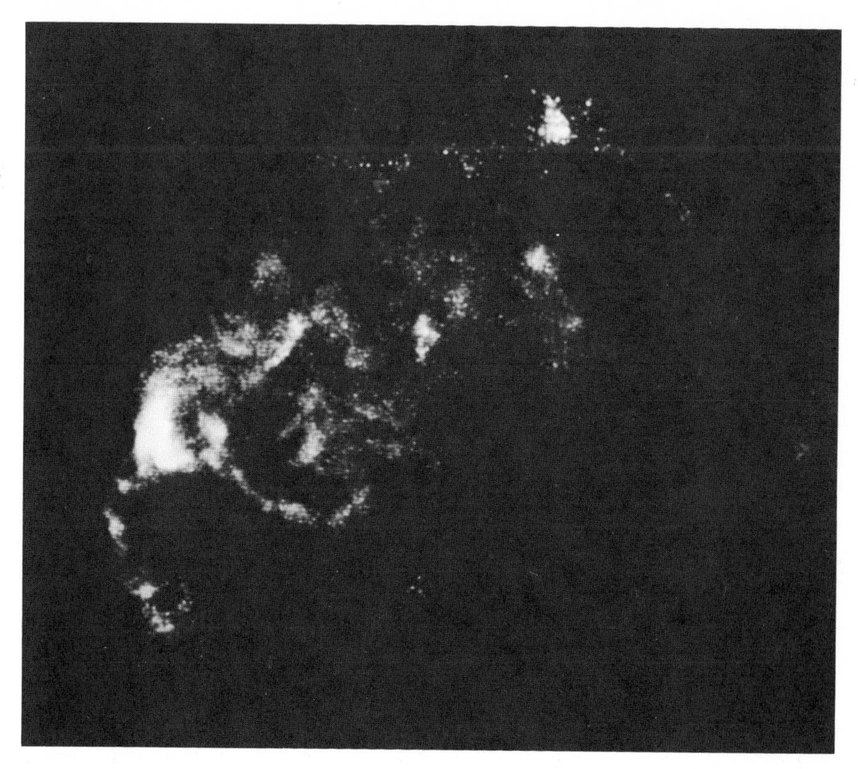

Figure 7.11
X-ray images of the Puppis A supernova remnant.

some of the major puzzles in x-ray astronomy. They are largely unidentified with optical counterparts, because of the high density of faint stars and the great visual obscuration in this part of the Milky Way. X-ray spectra measured with Einstein for a number of these objects provide much more precise values for the low-energy absorption, and hence the expected visual extinction, than was available before. The HRI has also been used to locate these sources on the sky to absolute position accuracies of about 2–3 arc-seconds. In most cases no optical counterparts are obvious, although several very faint stars are candidates for continued optical spectroscopy with very large optical telescopes.

In three galactic-bulge sources, however, very unlikely optical candidates are seen close to the x-ray position. These stars would appear to be unlikely counterparts for these apparently very luminous x-ray sources, because their optical spectra show them to be rather like the sun. However, if the sun were the optical counterpart (presumably the binary companion) of such a bright x-ray source, its appearance would be altered enormously; its atmosphere would be blasted by x rays and heated to a much bluer color, and its spectrum would contain several of the conspicuous emission lines now detected in the "normal" optical counterparts of strong galactic x-ray sources. This paradox may be resolved if these apparently solar-type stars prove to be not single stars but unresolved tight clumps of stars including an x-ray binary. These clumps of stars, perhaps no more than a dozen, might be the surviving remnant of an otherwise disrupted globular cluster in which the x-ray source was born. Before discussing this possibility further, it is valuable to describe the globular clusters themselves.

Globular clusters and x-ray bursters

Globular clusters are among the most interesting and well-studied objects in the galaxy. They are massive swarms, highly spherical in shape, of some 100,000 stars, all self-gravitating about a common center. Approximately 140 such star clusters are known in our galaxy; they are distributed spherically about the center of the galaxy. (Some galaxies contain thousands.) The Uhuru survey identified four globular clusters as the probable optical counterparts of galactic x-ray sources. These sources were known to be compact and to involve accretion onto a collapsed object, by virtue of their time variability, their x-ray spectra, and their high luminosities. (The distances to globular clusters can be measured optically, whereas distance estimates and thus absolute luminosities for most other compact galactic sources are ambiguous or very poorly constrained.) The temporal variations observed in the x-ray flux from globular cluster sources are irregular, and no periodicities, either pulsation or eclipsing, have been found despite intensive searches.

Still, the most extreme form of variability known in any galactic (or extragalactic) x-ray source was discovered from a source in a globular cluster: the phenomena of x-ray bursts. These are sudden increases (over about 1 second) in the x-ray brightness by a factor of typically 10–100 followed by a decline, usually smooth, to the original x-ray brightness over the next 5–10 seconds. An example of such a burst in the heavily obscured globular cluster Terzan 2 is shown in figure 7.12. This was the first of several x-ray bursts to be directly imaged by Einstein; all previous studies of bursters made since their discovery by the author in late 1975 had used detectors similar to those on Uhuru. Approximately 30 x-ray-burst sources have now been

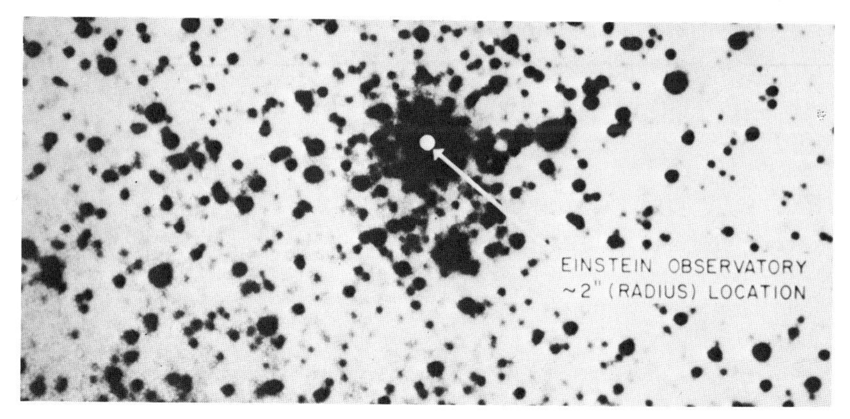

Figure 7.12
First direct x-ray image of an x-ray burst in progress and the location of its source within the globular cluster Terzan 2.

discovered, with a third of them in globular clusters, a third in the "field," and a third unidentified optically.

The bright x-ray sources and bursters in globular clusters were especially interesting since it was possible they were the manifestation of the massive black holes predicted as the result of the collapse of a cluster's stellar core. X rays would be produced by the accretion of gas (lost from stars in the cluster) onto the black hole. The alternative explanation for the luminous x-ray sources in globulars, and for the production of the x-ray bursts, was that they were compact binary systems with gas accreting from a low-mass star orbiting a neutron star. This second picture was supported primarily by detailed studies of the x-ray bursts, which were consistent with a model in which the bursts were thought to be due to a thermonuclear explosion of material (primarily helium) on the surface of the neutron star. Additional evidence that the bursters were low-mass binaries came from optical spectra of several bursters outside globular clusters. This result showed that a low-mass star was part of the system.

The Einstein Observatory finally confirmed the low-mass-binary hypothesis. The x-ray sources were "weighed" by using the high angular resolution of Einstein to locate the x-ray bursters in the clusters relative to the cluster centers to an accuracy of 1 or 2 arc-seconds. In theory, the more massive the x-ray source, the closer it should be to the exact center of the cluster. The new x-ray-source positions and the precise cluster centers were analyzed statistically to show that the most probable x-ray-source mass is only about 2 solar masses. (The mass must be less than 3 solar masses at the 90 percent confidence level and less than 5 solar masses with 99 percent confidence.) Thus, on the basis of their masses, the x-ray sources in globular clusters (and, by extension, the bursters both in and out of globular clusters) must be compact binary systems and not massive black holes.

The Einstein Observatory also turned up a large number of lower-luminosity x-ray sources in globular clusters. These too are almost certainly compact binary systems (although they have not been positioned on the cluster accurately enough to be weighed), but their lower luminosity suggests accretion onto a white dwarf rather than a neutron star. The results also suggest that there are a relatively large number of white dwarfs in cluster cores and that a substantial number of these are in binary systems. Collisions between the binary systems in a cluster core may then be expected to occur (over long time scales) and, in some cases, to expel one of the compact binaries from the cluster. This may be the origin of the x-ray bursters and the otherwise enigmatic bright galactic-bulge x-ray sources in the central region of the galaxy. The bursters and the galactic-bulge sources in the field are almost certainly compact binaries, like their brethren inside globular clusters. Perhaps globular clusters are the "stellar nurseries" for the formation and the subsequent ejection of these compact binary sources.

Some of the galactic-bulge sources may be compact binaries still in their "parent" clusters but with their surrounding clusters almost entirely stripped away by repeated passages through the plane of the galaxy. Clusters will be torn apart by gravitational tidal forces of the galaxy if they spend most of their time near the galactic center or in the galactic plane. This picture would also explain the interpretation given above for the apparently sunlike stars identified with several bright galactic-bulge x-ray sources that are also bursters. They might be the tight binary system surviving in an otherwise stripped globular cluster. This possibility can be tested by future optical observations with the Space Telescope, for it will have sufficient angular resolution to determine whether these optical objects are very compact aggregates of stars or simply single stars.

Probing the center of the galaxy

The nucleus of the Milky Way is of special interest because it is the center of rotation and mass of our entire galaxy, but the physical conditions there are poorly understood. Whereas visible light is totally absorbed by the intervening gas and dust, radio waves, infrared emission, x rays, and gamma rays are all detectable from the very innermost region. Indeed, studies of the velocities of gas clouds detected in this

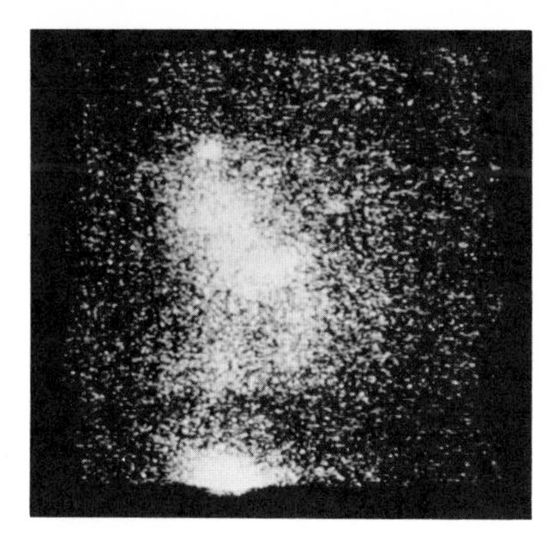

Figure 7.13
X-ray image of the center of our galaxy. The bright source at the bottom of the picture is a foreground source.

region by their infrared emission suggest that there may be a massive black hole—perhaps a million times more massive than the sun—lurking at the heart of our galaxy. Such black holes may be typical of the nuclei of most galaxies, since our current understanding of those galaxies with violently active nuclei, which include the quasars, increasingly points to their being powered by accretion onto even bigger (100 million solar masses) black holes.

Once again, early Uhuru observations had indicated that the nucleus of our galaxy was a rather weak x-ray source. However, the situation was unclear, because complex emission was detected over a region of approximately 0.5°. The Einstein images have finally resolved the galactic center, as shown in figure 7.13. The galactic center is the brightest spot in the otherwise diffuse emission that covers much of the image. Within this diffuse emission, some dozen much fainter pointlike sources are also visible. Two of these may be identified with foreground stars, but the bulk are probably at the 25,000-light-year distance of the galactic center itself. Both the diffuse emission and the weak point sources embedded in it are asymmetric about the galactic center in the same way as both the radio and infrared emission from this complex. This suggests a common association between the gas and dust, as traced by the radio and infrared emission, and the x-ray sources. The diffuse x-ray emission might, therefore, be due to complexes of hot stars and/or their resulting supernova remnants. And the faint point sources could be their collapsed remnants accreting gas from the interstellar clouds.

The galactic nucleus itself is relatively faint at the low x-ray energies detected by the Einstein Observatory. Its luminosity is comparable to or even lower than those of typical x-ray binary systems, and thus much less than the x-ray luminosity seen from the nuclei of active galaxies and quasars. If the source is due to accretion onto a massive black hole, then the accretion rate must be very low. Perhaps the nucleus of our galaxy, like those of most galaxies, is now dormant. It may have been luminous and active in the distant past, when the active galaxies or quasars were apparently much more luminous.

SS433: A miniature active galaxy

Perhaps the most exotic member of the x-ray "zoo" in our galaxy is SS433, which was considered simply another weak and undistinguished x-ray source discovered in the Uhuru survey until it was identified with a very peculiar optical star. The spectrum of emission lines (mostly of hydrogen and helium) from the star revealed that most of the radiating gas was moving in two oppositely directed beams, each with a velocity of more than a quarter the speed of light. Such velocities (not to mention such beamed geometries) were unprecedented for any object previously known within our galaxy, but they were reminiscent of the emission jets often seen emerging from active galactic nuclei. However, the physical reality of SS433's jets was not totally established from the optical data; the jets were merely an apparently simple and predictively successful model for explaining the observations.

The reality of the beams in SS433 was only revealed when the first Einstein x-ray images of this remarkable object were obtained in 1979. The composite x-ray image recorded by several different Einstein observations is shown in figure 7.14. X rays from giant diffuse lobes are detected out to more than 0.5° from the bright central point source. (Only this central point source was detected by Uhuru.) The diffuse lobes produce only about 10 percent of the total x-ray emission. Moreover, the x-ray lobes are at just the angle on the sky predicted by the model for the optical beams, which are themselves unresolved (in optical photographs) from the central point sources. Thus, the central source, also now known to be a binary system, is indeed expelling enormous quantities of matter in two beams, which first glow in visible light and then, much later and farther out (at a distance of some 30 light years from the central object), somehow "heat up" to much higher temperatures and radiate x rays.

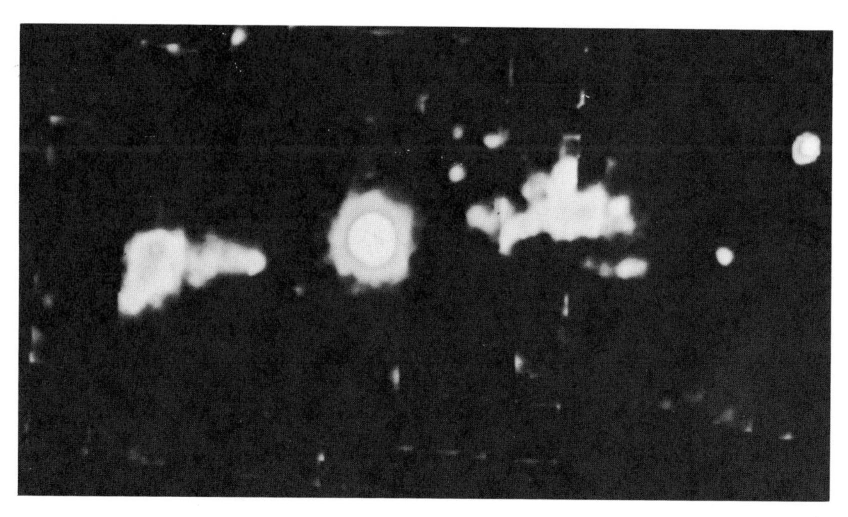

Figure 7.14
X-ray image of the peculiar galactic x-ray source SS433, showing its striking double lobe structure, which is similar to that often observed in active galactic nuclei.

An x-ray overview of the galaxy

The galaxy becomes a rich and varied sight when viewed in the x-ray range rather than the visible-light range. The very brightest sources are pointlike and are grouped near the center and along the galactic plane. These are all time-variable and appear to the x-ray eye to be flickering in brightness, with typical variations of a factor of 2. Occasionally, very bright and fast flashes—x-ray bursts—are seen. These sources have an intimate connection with those in globular clusters and are probably all compact binary systems composed of a low-mass star orbiting a neutron star. It is these galactic-bulge and globular-cluster x-ray sources that are so prominent in the x-ray image (figure 7.2) of the Andromeda nebula. The bulge sources in our galaxy would appear less concentrated about the center to any x-ray astronomers viewing us from Andromeda.

Farther out from the galactic bulge and more concentrated on the plane of the galaxy are the bright x-ray pulsars such as Cen X-3. These sources are also neutron stars in binary systems, but they differ from the galactic-bulge sources in several crucial respects. First, the neutron stars are apparently much more highly magnetized: that is, matter accretes only onto limited portions of them. The rotation of these "hot spots" produces the characteristic x-ray pulses. The normal stellar companions in these systems are also different. They are usually massive early-type stars in which mass loss is occurring naturally in the form of winds. In comparison with the galactic-bulge sources (which, because of their association with globular clusters and their central position in the galaxy, must be products of very ancient stellar systems), the x-ray pulsars in massive binaries must be very young. This is so simply because the massive normal stellar companions in these systems do not live very long themselves. Their lifetimes are typically only 10 million years. Then they, too, may explode and leave behind neutron stars or black-hole remnants.

The vast majority of the bright point sources of x rays in our galaxy appear, then, to be neutron stars in binary systems. In only two cases is there sufficient evidence to suggest that the compact object is a black hole rather than a neutron star. And in only one of these cases (Cygnus X-1) is the evidence really compelling. The other case (GX339-4), for which the optical counterpart was only recently found by the author, requires considerable further observational study. Apparently black holes are rare as the end products of stellar evolution. Perhaps, even in the most massive stars, considerable mass loss occurs before and during the supernova event, so that the remnant stellar cores are almost always less than the lower limit for black holes (about 3 solar masses).

Supernova remnants themselves are spectacular objects in the x-ray range. They glow in x rays for perhaps 50,000 years and then fade away. Some, but apparently few, have central pulsars to supply them with energy in the form of highly energetic particles (cosmic rays). These particles, in turn, give off continuum radiation across the entire electromagnetic spectrum, from the radio band through the x-ray and gamma-ray bands. Most supernova remnants contain at least a low-energy spectrum of particles, which can produce synchrotron radiation in the radio band. The Crab nebula exemplifies the relatively unusual type of SNR with an energetic central pulsar powerhouse producing an evenly filled diffuse x-ray emission by the synchrotron process. More typical SNRs are shell-like or wispy structures in the x-ray range. The x rays are produced by hot filaments of gas swept up in the supernova blast and heated by shock waves to temperatures of tens of millions of degrees. The spectrum of this thermal x-ray emission reveals details not previously available on the ionization and the composition of the remnant gas.

The "normal" stars that eventually evolve to produce the x-ray sources described above are themselves detectable in the x-ray range. The x-ray images from the Einstein Observatory have shown that not only the extreme conditions (stellar flares) but also the normal quiescent emission are detectable in the x-ray range. Most stars appear to have hot coronas, or thin outermost atmospheres, emitting x rays. This discovery has had a major impact on theories of the formation and the heating of stellar coronas, including that of our sun, and it now seems more certain than ever that dissipation of magnetic field energy is involved.

Perhaps the most unexpected discovery is the relatively bright x-ray emission, indicative of high-temperature gas, from even the apparently coolest and least luminous stars. These are the so-called red dwarf stars, with masses typically only 10 percent that of the sun and visual luminosities only 0.1–1 percent that of the sun. Yet their quiescent x-ray emission rivals or exceeds that of the sun and is responsible for almost 10 percent of their total energy output. This surprising discovery made with the Einstein Observatory, like many others made both in x-ray astronomy and in astronomy as a whole, was totally unanticipated and will spur new theoretical developments. Since red dwarfs are, as the lowest-mass stars, those most commonly found, they are the dominant stellar population in the galaxy and their total contribution to the overall x-ray luminosity of the galaxy is significant—it is probably about equal to the x-ray luminosity in the total of the brightest accretion sources, the compact x-ray binaries. Thus, a very sensitive x-ray image of our entire galaxy (from Andromeda, say) would show it to be glowing diffusely in x rays from the sum total of its stars.

The galaxy's total x-ray luminosity is probably at least 10 percent of its total energy output, most of which is optical and infrared radiation. This is a lower limit, however, since we have still not explored the galaxy with high sensitivity at either very low x-ray energies (where absorption by galactic dust is important) or very high x-ray energies (where detectors have not yet achieved sufficient sensitivity or imaging capability). Whereas the galaxy quickly becomes opaque at very low x-ray energies and will therefore probably never be mapped in full detail, high-energy x-ray astronomy holds great promise for the future. New detectors with both an imaging capability and much higher sensitivity are being developed for observations from space. These, together with the Advanced X-ray Astrophysics Facility (AXAF), planned as a permanent x-ray telescope 100 times more sensitive than Einstein, will reveal an increasingly clear, and perhaps still unanticipated, x-ray portrait of our galaxy.

Reading

Giacconi, Riccardo. "Progress in X-Ray Astronomy." *Physics Today* 26 (May 1973): 38–47.

Giacconi, Riccardo et al. "The Einstein (HEAO-2) X-Ray Observatory." *Astrophysical Journal* 230 (1979): 540–550.

Giacconi, Riccardo. "The Einstein X-Ray Observatory." *Scientific American* 242 (1980): 80–102.

Harnden, F. R., Jr. "X-Ray Telescope in Space Reveals Unseen Universe." *Smithsonian* 11 (September 1980): 110–114.

Lewin, Walter H. G. "The Mystery of the X-Ray Burst Sources." In *Revealing the Universe*, James Cornell and Alan Lightman, eds. Cambridge, Mass.: MIT Press, 1982.

NASA Public Affairs Division. High Energy Astronomy Observatory. Report EP-167. Washington, D.C: Government Printing Office, 1981.

Overbye, Dennis. "The X-Ray Eyes of Einstein." *Sky and Telescope* 57 (June 1979): 527–534.

Ruffini, Remo, and John A. Wheeler. "Introducing the Black Hole." *Physics Today* 24 (January 1971): 30–39.

Sullivan, Walter. *Black Holes.* New York: Doubleday, 1979.

8

X Rays Beyond the Milky Way

Paul Gorenstein

To the professional astronomer and the layman, the questions that evoke the most interest are the recurring ones about the universe as a whole. What does the universe look like on a large scale? Does it contain galaxies very different from our own (the Milky Way) and its neighbors? Has the universe changed as it has aged, and will it appear different in the future? What will be the final fate of the universe? To answer these questions, we must search far beyond the Milky Way.

Almost everyone has heard that, because light travels with a finite velocity, objects observed at large distances are seen as they were in the past. However, only at distances beyond 3 billion light years is the light travel time sufficiently large for the universe to have undergone substantial changes. Distances of this magnitude and beyond would be called cosmological distance. The major goal of extragalactic research, then, is to observe the universe in various phases of time, to reconstruct its evolutionary history, and to predict its future development.

The variety of data available to astronomers has increased remarkably during the past 20 years. We now can examine the universe through new windows of the electromagnetic spectrum (the radio, gamma, and x-ray bands) and also at the infrared and ultraviolet extensions of the more traditional window of optical astronomy. The view through these new windows is not only helping to provide answers to the fundamental questions cited above but is also altering our perception of the universe. The effect has been so profound that one could truly say that the twentieth century's second revolution in astronomy is occurring.

The first revolution, which coincided approximately with the century's first three decades, was the recognition of the existence of external galaxies. The galaxies were determined to be receding from each other at a speed that, on the average, is proportional to their distance of separation. This picture is consistent with the idea that the matter from which all galaxies are formed was initially at the same point some 10–20 billion years ago. A giant explosion known as the "big bang" initiated the outward motion of the matter. "Majestic" and "immutable" are adjectives that were widely used several years ago to describe the universe. However, this picture is still incomplete.

The second astronomical revolution has been the realization that the universe is not static but is a dynamic entity constantly experiencing dramatic changes. Indeed, episodes of violence have occurred often in the universe. Very-high-energy particles, extremely hot gas, high magnetic fields, strong gravitational fields, and matter that cannot be seen in visible light are factors accounting for the extreme conditions that characterize the early universe and are still seen in an attenuated form in certain nearby objects. In this regard, x-ray measurements are crucial, for they are the most intimately associated with these extreme conditions.

X rays are produced in the following ways. High-energy particles follow spiral paths when they encounter magnetic fields. Their deflection is accompanied by the emission of electromagnetic radiation along their tangential direction in a process known as synchrotron radiation. The highest-energy electrons radiate in the x-ray band. In addition, wherever electrons of moderately high energy are found in large numbers, their presence is notable as a result of their collision with photons at radio wavelengths. In a process that is known as the "inverse Compton effect," a high-energy electron elevates the energy of a photon from the radio band to the x-ray band. The source of the radio photons could be the pervasive background of microwave radiation. Many astronomers believe that background to be the present, highly red-shifted state of radiation produced in the "big bang" of creation some 20 billion years ago. Or, the radio photons could be produced by the electrons themselves as a result of the synchrotron process. In either case, the collisions of high-energy electrons with low-energy photons in the radio band will result in large clouds of x-ray emission from the sites where moderately high-energy particles are concentrated. A third process for producing x rays also figures prominently in studying the regions beyond the Milky Way. This is the process of thermal radiation from gas having a temperature in the range of millions to hundreds of millions of degrees. X rays are the characteristic radiation for this temperature range, just as visible light is the characteristic thermal radiation from the surfaces of stars that are at temperatures of several thousands of degrees. The relatively recent detection of the presence of this hot gas in intergalactic space has very profound implications for our understanding of the fundamental questions posed at the beginning of this chapter.

A common element in all of the extragalactic x-ray-production processes is that all involve forms of matter invisible to optical telescopes. Moreover, each of the various kinds of extragalactic x-ray emission is associated with a form of dark matter. The dark matter falls into two extremes: matter that is very dense and matter that is very diffuse in comparison with the sources of visible light, namely stars. Compact dark matter is the powerhouse that may explain the high luminosity of quasars and other extraordinarily active galaxies. Diffuse dark matter, which is less well understood, may be the most abundant form of material that exists in the universe and could very well be the decisive factor in determining its ultimate fate.

It is important to emphasize that when we refer to *extragalactic* x-ray emission we mean phenomena on a scale much larger than the individual x-ray sources within our galaxy. The brightest galactic x-ray sources are associated with individual stars that have run through their normal life cycles and entered either extremely condensed or explosive phases. A normal galaxy, such as the Milky Way, may contain from 100 to 1,000 of these. Extragalactic x-ray sources are anywhere from 10,000 to 100 million times more powerful than the brightest galactic sources. They are associated with phenomena that involve galaxies and clusters of galaxies. Normal galaxies, such as the neighbors of the Milky Way, are not examples of what we call extragalactic x-ray sources.

As Grindlay explains, it is necessary to get above the atmosphere to observe cosmic x rays. This requires either satellites or rockets to lift the instrumentation beyond an altitude of 100 miles. Satellite experiments are rather expensive, and opportunities to launch satellites are limited. Thus, the number of satellite experiments worldwide has been rather small. The largest and most recent x-ray-astronomy satellite was the Einstein Observatory, which provided a great deal of new information on x rays beyond the Milky Way. Most of the material and illustrations in this chapter are based on the results obtained from it. The Einstein Observatory is no longer operating, having reentered the Earth's atmosphere in March 1982, but analysis of the data that it telemetered back to Earth will occupy astronomers for several years to come.

The extragalactic x-ray sky

It is interesting to compare the appearance of the sky in the x-ray and the visible-light ranges. The specificity of x-ray measurements for extragalactic research becomes apparent in this comparison. Figure 8.1 shows a region measuring about one square degree obtained as part of a survey of the northern sky by the 48-inch Schmitt telescope on Mount Palomar. This field is about 32° north of the midplane of the Milky Way, and it is shown here because it is centered on a cluster of galaxies cataloged as Abell 2256. Clusters of galaxies will be described in more detail later; the point here is that those objects that are interesting for extragalactic studies do not immediately stand out in the photo. Like most visible-light photographs, this one is dominated by faint starlike images. Indeed, most of the pointlike images are faint stars within our own galaxy. A few other (apparently stellar) images are actually the bright nuclei of distant active galaxies, which include quasars. Other types of measurements, namely spectra, are required to distinguish between the quasars and stars. The small diffuse images are those of external galaxies; the central galaxies belong to the cluster A2256.

The same region of the sky is shown in the x-ray image in figure 8.2. The image is derived by joining four contiguous exposures, each 12,000 seconds long, obtained with the Imaging Proportional Counter of the Einstein Observatory. In contrast with the optical photograph, unusual extragalactic objects stand out immediately in the x-ray image. Although there are far fewer objects in the x-ray picture, most of the point sources are now extragalactic in origin—distant quasars or active galaxies, with only one or two stars among them. The cluster of galaxies A2256 stands out vividly in this field as a large diffuse image that occupies one-fifth of the field in each dimension. The diffuse image is due to thermal radiation from a gas at a temperature of 100 million degrees. The presence of iron lines in the x-ray spectrum of the gas suggests that the gas was swept out from stars within the galaxies into the medium between galaxies. A diffuse x-ray image is an identifying characteristic of clusters of galaxies such as A2256. Thus, clusters of galaxies, including ones too distant to be identified by any optical means, can be identified immediately.

In the sections that follow, both types of x-ray sources will be described in more detail. The pointlike x-ray images associated with active galaxies are cases where dark matter of high density is responsible for the emission. They include some of the most powerful objects known to exist in the universe. The extended, or diffuse, images of

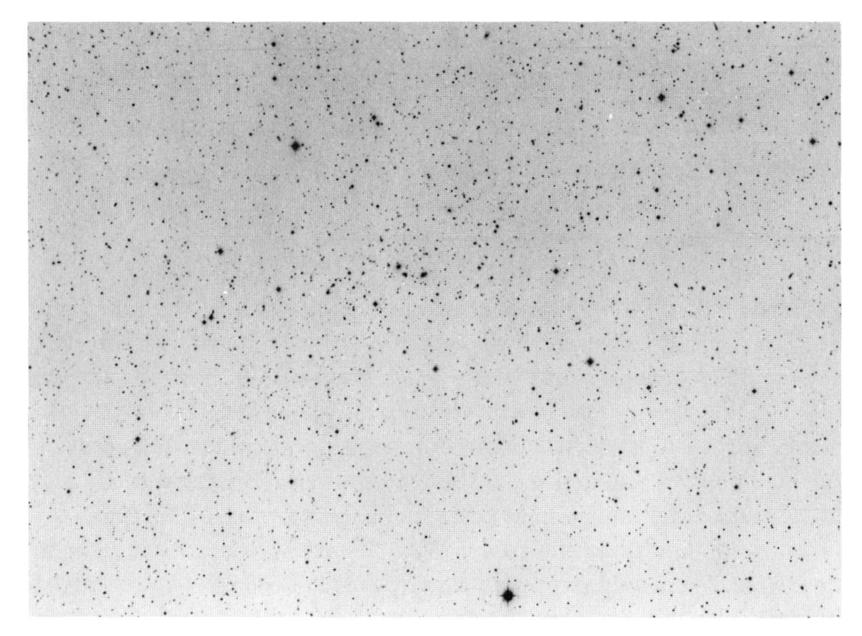

Figure 8.1
Optical photograph of region of sky containing the cluster of galaxies Abell 2256, taken with the 48-inch Schmidt telescope on Mount Palomar. (Hale Observatory photo)

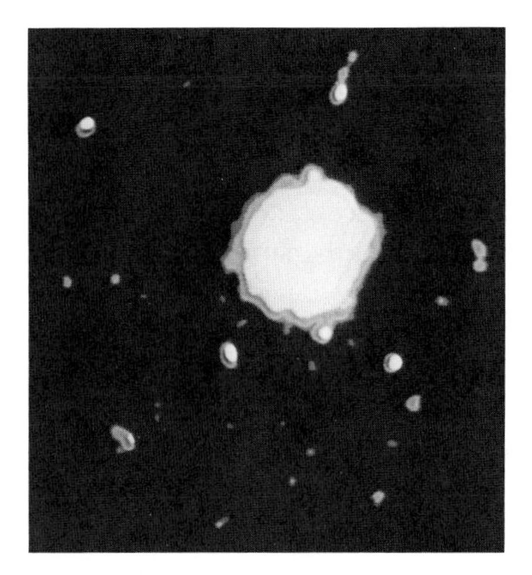

Figure 8.2
X-ray image of the region of sky shown in figure 8.1, obtained from the Imaging Proportional Counter of the Einstein Observatory. The diffuse image is the cluster; the pointlike images are mostly distant active galaxies. (Smithsonian Astrophysical Observatory photo)

clusters are at the opposite extreme. Here a tenuous gas with a temperature approaching 100 million degrees is subject to the gravitational influence of another (more massive) component of matter that cannot be seen. This dark component may be the most abundant form of matter in the universe.

Normal galaxies

In x-ray activity, the galaxies in our immediate cosmic neighborhood are not radically different from our own galaxy. Thus, although they are formally outside of the Milky Way, they are not prime examples of extragalactic sources. However, they do play a very important role in helping us to understand the differences and similarities between normal galaxies and active ones. In the constellation Andromeda and at a distance of 2 million light years, the galaxy known as M31 appears to be quite similar to the Milky Way. It is shown in visible light in figure 8.3. We would expect an x-ray picture of M31 to be dominated by a collection of individual sources much like the brightest compact binaries and the young supernova remnants within our galaxy. Our expectations are fulfilled. However, by observing M31, we do have the opportunity to see the entire distribution of x-ray sources in a normal galaxy with only a few pointings of the Einstein Observatory's telescope. Two views are shown. Figure 8.4 is the entire galaxy as seen with the wide-field but low-resolution Imaging Proportional Counter of the Einstein Observatory. The central bulge of figure 8.4 is resolved into an array of sources in a closeup of the central region (figure 8.5) taken with the High Resolution Imager.

Although studies of these sources are important primarily for understanding more about our own galactic sources (and normal galaxies in general), they also bear some relation to our main concern in this chapter: extragalactic x-ray sources. The connection is through a single x-ray source at the center of M31. Although this central source is no brighter than average, the origin of its emission may be totally different in nature than the others. It radiates at a level of 10^{37} ergs per second, which means that it radiates 10,000 times as much energy as the sun, with the sun emitting almost all of its power in its principal band, visible light. (It is useful to keep in mind the figure of 10^{37} ergs per second for later comparisons with more active galaxies.) The center of our galaxy contains an x-ray source, too. At 10^{35} ergs per second, it is much less luminous than the source in M31. However, it does radiate with stronger, but variable, intensity in gamma rays.

Figure 8.3
Optical photograph of our neighbor galaxy, M31, known historically as the "great Andromeda nebula." (Lick Observatory photo)

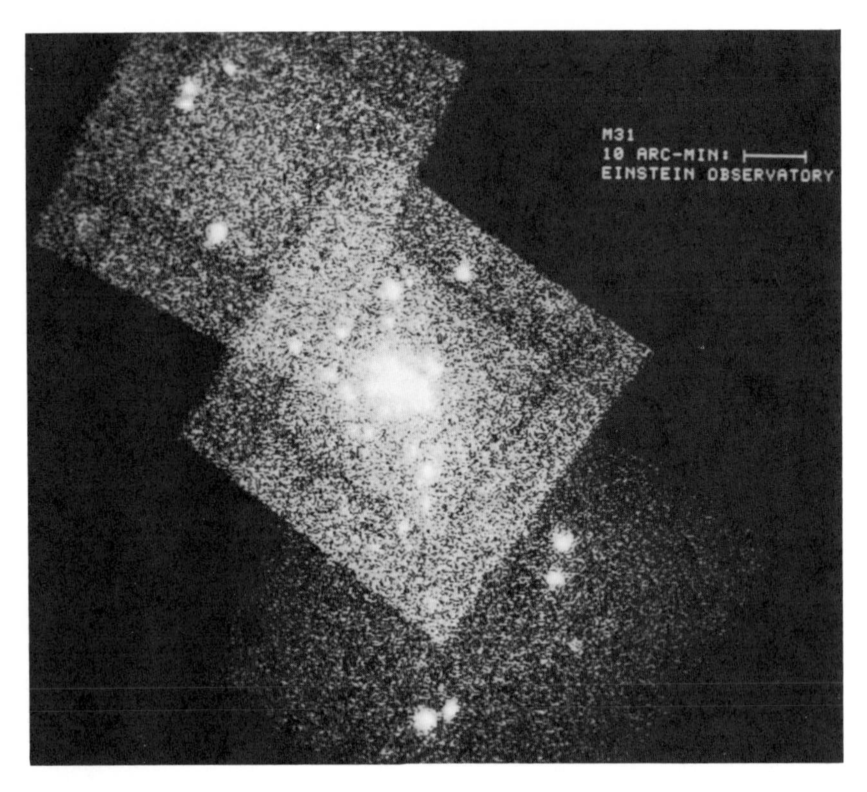

Figure 8.4
X-ray photograph of M31 obtained with the Imaging Proportional Counter of the Einstein Observatory. Individual galactic-type sources and a central bulge dominate the photograph. (Smithsonian Astrophysical Observatory photo)

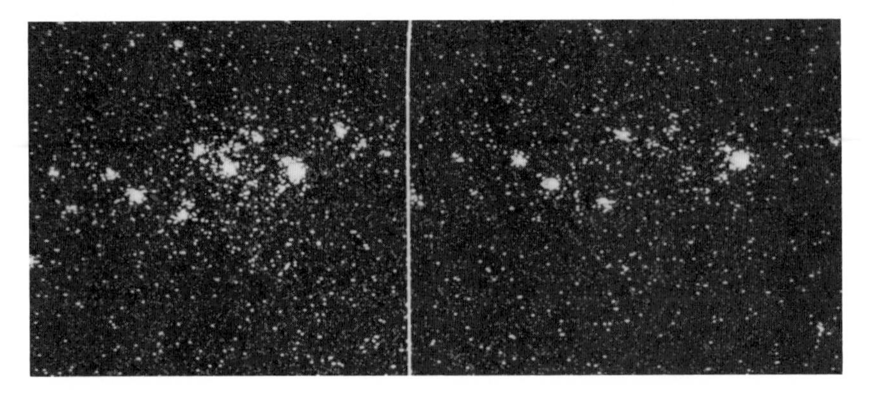

Figure 8.5
In this photograph, sources in the central bulge of figure 8.4 are resolved with the High Resolution Imager of the Einstein Observatory. A comparison of two observations made 6 months apart reveals intensity variations in several of the sources. (Smithsonian Astrophysical Observatory photo)

As we look a few times beyond the distance of M31 into other normal galaxies, we uncover occasional examples of individual or compact binary systems that radiate with much more intensity than any within either our own galaxy or M31. Frequently, they are found in galaxies with a peculiar morphology or in galaxies that have been disturbed by a close interaction with a nearby galaxy. It is believed that such disturbances lead to the creation of groups of new stars. The more massive of these stars evolve rapidly, undergo supernova explosions, and form compact binary systems. Occasionally one such system will be extremely active. Although some of these sources are very interesting subjects in their own right, the x-ray emission phenomena are still essentially similar to what is occurring within our own galaxy.

Active galaxies

At a distance of 25 million light years we detect one of the first examples of what could truly be called an extragalactic x-ray source: the galaxy NGC 5128. This galaxy is one of the brightest radio sources and is known as Centaurus A to radio astronomers. At the distance to Cen A, individual sources within the galaxy are too faint to be seen. But 25 million light years is still fairly close in comparison with cosmological distances, and the universe probably has not undergone any significant changes in the 25 million years that it took for light (or x rays) from Cen A to reach us. Essentially, Cen A is our contemporary in galactic evolution. However, it may be a relatively rare surviving example of a phenomena that was more common in earlier epochs of the universe's history. Observation of these comparatively near active galaxies is extremely important. They are close enough to provide an opportunity for detailed study of phenomena that are manifested more extremely in more distant objects but would be too faint and difficult to resolve at that larger distance.

In an x-ray photograph of Cen A (figure 8.6) important features appear that go beyond what is seen in normal galaxies. For example, a bright spot at its center has an x-ray luminosity exceeding 10^{42} ergs per second, 100,000 times that of the center of M31. Its output of x rays has been observed to more than double and fall again within the course of the year. Another feature is a jet of x-ray emission emanating from the nucleus of NGC 5128. The jet can be ascribed to a beam of particles accelerated by some mechanism that receives its energy from the center.

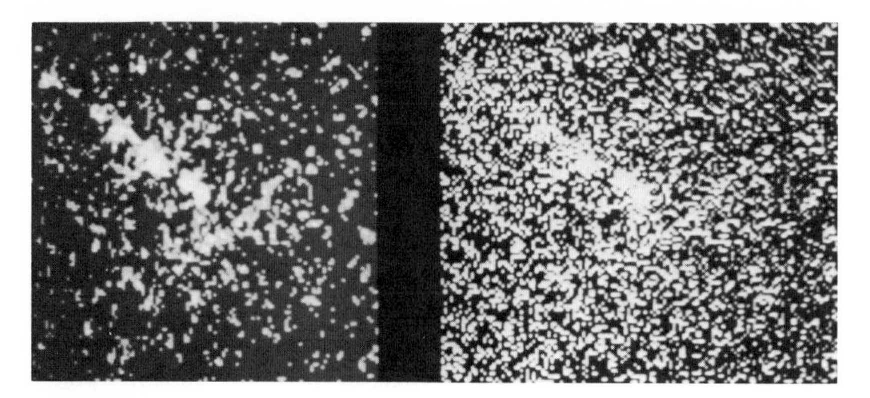

Figure 8.6
X-ray image of the radio galaxy Centaurus A made with the High Resolution Imager. The right panel is a raw data image; the left panel is the result of computer-assisted image enhancement. A jet of x-ray emission emanates from the center of the galaxy toward the northeast. (Smithsonian Astrophysical Observatory photo)

Another comparatively nearby active galaxy is the giant elliptical galaxy known as M87 to optical astronomers and as Virgo A to radio astronomers. Several important aspects of M87 are revealed by x-ray measurements, but at this point I will concentrate on the central region. A short exposure in visible light (figure 8.7) shows a jet emanating from the center of M87. The polarization of the light indicates that it is produced by synchrotron radiation from high-energy electrons moving in the magnetic field of the galaxy. The nucleus and the brightest points along the jet are also seen in x-ray observations. Their intensity suggests that synchrotron radiation is responsible. An x-ray view of a larger region in the center of M87 is shown in figure 8.8. The nucleus and the jet are barely resolved in this view, which shows larger-scale features. Jetlike lobes of emission appear to the left (east) of the center and to the southwest of the center, probably produced by the inverse Compton effect, whereby high-energy electrons collide with radio photons. The implications of detecting synchrotron emission for the jet in the x-ray band are profound. To be seen at a distance of 50 million light years, the linear size of the M87 jet must be about 4,000 light years. However, high-energy electrons cannot radiate x rays by the synchrotron process for a period of more than a few years. The detection of an x-ray jet larger than a few light years means that the energy of the electrons is being maintained by a reacceleration process that must take place along their path. The basic energy supply of the beam must be something other than electrons. Thus, the picture

Figure 8.7
A short-exposure photograph of the giant elliptical galaxy M87 obtained by the Lick Observatory. Only the central region of the galaxy and a bright jet pointing toward the northwest appear. The length of the jet is about 0.25 arc-minute. (Lick Observatory photo)

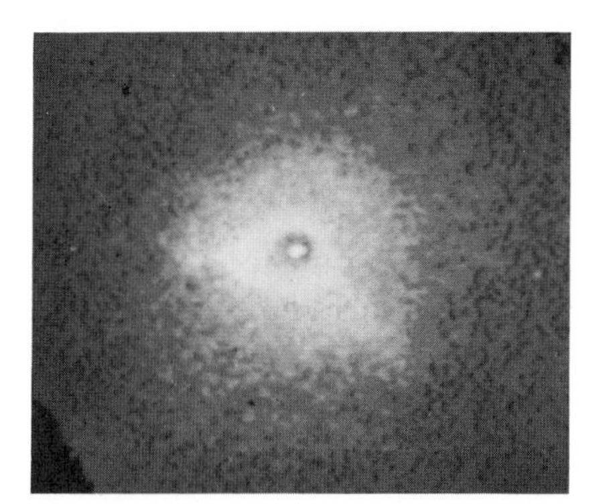

Figure 8.8
X-ray photograph showing a larger region in M87, obtained by the High Resolution Imager. (Smithsonian Astrophysical Observatory photo)

of M87's central region—and, by induction, those of all active galaxies—is the following: Beams of protons (or perhaps uncharged hydrogen atoms) are shot out of the nucleus into a region that contains clouds and magnetic fields. By processes that are rather complex, whenever the protons collide with interstellar clouds, electrons are accelerated to high energies. These electrons spiral along magnetic field lines. This occurs at several points along their trajectory across the galaxy. At each collision point with a cloud, we detect the synchrotron radiation from electrons spiraling in the magnetic field. Most of the original energy in the particle beam remains, and as the process repeats the result is a jet.

There is yet another important result pertaining to M87. Optical astronomers have determined that near the center of the galaxy both the clustering of stars and their distribution of velocities suggest that the stars are under the influence of an intense gravitation field. Such a field could arise from a compact object with a very large mass, perhaps as much as that of 5 billion suns. If this is correct, then this massive central object in M87 could be a giant black hole. Since a black hole cannot be observed directly, one must observe a black hole through its gravitational effects on external objects, such as the stars in M87 that surround it. Another indirect method of observation is through measurement of the radiation that is given off by the streams of infalling matter heated intensely during mutual collisions in the final orbits before disappearing into the black hole. The power source is the conversion of potential energy into heat or kinetic energy of outgoing particles. As matter falls into the intense gravitational field before disappearing into the black hole, the temperature and the kinetic energy of the particles should be high enough for x rays to be the predominant form of radiation. Fluctuations in the streaming or in the final revolutions around the black hole may be reflected as temporal variations in the x-ray flux. With more matter available in the vicinity of the black hole, more is likely to fall toward it, and the output of x rays should increase, at least up to a point where the density becomes too great for even x rays to escape.

Another type of active galaxies, known as Seyfert galaxies after the astronomer who first described them some 40 years ago, are strong pointlike x-ray sources. Although their nuclei are not always strong radio emitters, they are very bright in both the visible-light and infrared ranges. The spectrum of the nucleus of a Seyfert galaxy is not like that of ordinary starlight; rather, it shows manifestations of a region highly excited by an additional source of energy. The remainder of the galaxy is usually of a spiral type, not unlike M31 and the Milky

Way. Seyfert galaxies are an important link between normal spiral galaxies and the more intense forms of active galaxies found at very large distances.

The most luminous objects in the universe are found at large distances from the Milky Way. The first of these to be identified optically were found on the basis of their radio emissions and were called quasistellar radio sources, or quasars. Because their bright nuclei appear starlike in visible light, they were later given the more general name quasistellar objects, or QSOs. Subsequently, many QSOs were discovered that were not radio emitters. However, all QSOs are strong x-ray sources. In fact, most radiate nearly as much energy in x rays as in visible light, and a significant percentage emit most of their radiation in the x-ray band. Strong red shifts in their spectral lines also suggest that QSOs are at very large distances. While there is no unanimity among astronomers on this interpretation, most believe that QSOs are at cosmological distances and that their luminosity is very large. Increasing evidence, supported by recent x-ray measurements, indicates that QSOs are younger and more luminous examples of Seyfert galaxies and other active galaxies. (Of course, if a normal-appearing spiral or elliptical galaxy did surround the bright nucleus of a QSO, it would be too faint to detect at cosmological distances.) The volume density of QSOs increases with distance out to a point that corresponds to a lookback time of 10–20 billion years. The relative lack of QSOs at close distances (that is, in the present epoch) means that they are an evolving population. These objects seem to have first appeared about 10–20 billion years ago and, for the most part, to have either disappeared or evolved into a more common type of galaxy some time after 5 billion years ago—about the time of formation of the solar system.

As mentioned earlier, an optical photograph of the sky is dominated by the visible light of normal stars, either as individual members of our galaxy or as entire external galaxies. QSOs and other active galaxies account for only a very small percentage of the optical images. Thus, identifying QSOs on the basis of visible-light measurements is a very arduous process requiring a great deal of care and cross-checking with spectral data. On the other hand, x-ray sources are much more rare, and x-ray emission is an identifier of all active galaxies. Indeed, a very large percentage of those sources not obviously associated with nearby stars turn out to be either QSOs or active galaxies. A typical six-hour exposure of a 1° field with the Einstein Observatory resulted in the detection of about ten new sources, most of which were QSOs.

One salient feature of the x-ray emission of active galaxies is the occurrence of time variability on a scale of hours or faster. This is highly significant, for it sets a fundamental limit on the size of the emitting region. According to well-established principles of physics, the size of an emitting region that varies can be no larger than the speed of light multiplied by the time in which the variation occurred. Thus, the observation of a significant variation in one hour implies that the radius is smaller than about eight times the distance from the Earth to the sun—a very small distance in comparison with the dimensions of a galaxy or even the typical distance between stars. Thus, because of their specificity to active galaxies, x-ray measurements by the Einstein Observatory and other x-ray-astronomy satellites represent an important body of new information for studying QSOs. There is an opportunity for studying their evolution and for understanding their connection to objects of the current epoch we see at relatively near distances.

The following list summarizes x-ray observations relevant to active galaxies.

Essentially all active galactic nuclei (whether radio emitters or not) contain pointlike x-ray sources, and the x-ray flux is usually a significant percentage of the total power.

The average x-ray luminosity increases progressively from Seyfert galaxies to nearby QSOs to distant QSOs. Indeed, the continuity of their behavior in the x-ray band is rather pronounced, more so than in the optical band.

Rapid x-ray time variations indicate that the power source is compact (no larger than a few light hours).

For the most powerful galactic x-ray sources, the source of the power is known to be the conversion of gravitational potential energy into kinetic energy by the accretion of matter onto a compact object (either a neutron star or a black hole). One compact binary system, SS433, exhibits jetlike x-ray features due to particle beams that are reminiscent (at least superficially) of the radio jets seen in radio galaxies and some quasars.

On the basis of optical measurements, there is evidence for a massive object at the center of the radio galaxy M87 that emits very little or no light. However, we do detect x rays from the center of M87.

Pointlike x-ray sources of rather low intensity are seen in normal galaxies such as our neighbor M31 and at the center of our own galaxy.

Although the picture is incomplete and one must make some assumptions, these observations allow us to construct a model for QSOs and how they evolve with time. In so doing, however, one must assume that certain observations made for a specific active galaxy will apply to all. The large mass of the object at the center of M87 that optical astronomers have proposed, plus the small size suggested by x-ray time variations over an interval of an hour seen in other objects, are consistent with the presence of a massive black hole. Thus, the model for the power source of QSOs and other active galaxies is accretion of matter onto a central black hole. Gravitational potential energy of matter falling into this black hole is transformed in part into radiation and atomic excitation and into the kinetic energy of particles expelled outward. Outgoing particles will undergo collisions that lead to the acceleration of electrons. Relatively close to the black hole, the electrons are of highest energy and radiate x rays by the synchrotron process. At larger radii from the black hole, the radiation is predominantly in the optical band. At still larger radii, where the average electron kinetic energy has fallen considerably, radiation is primarily at the radio wavelengths. The various regions and their sizes are illustrated in figure 8.9.

Variations in power output from one active galaxy to another are explainable as individual differences in the sizes of the black holes and in the amount of matter available to be accreted upon them. QSOs, Seyfert galaxies, and normal spiral galaxies all represent different stages or conditions of what are essentially similar objects; a massive black hole is at the center of each. (A similar connection applies to the radio emitters: quasars and radio galaxies.) The difference in the luminosity of the three objects is explainable by conditions around the central black hole. In young objects, namely the QSOs, the reservoir of matter or fuel in the immediate vicinity of the central black hole is greatest and conditions favor a high efficiency for converting gravitational potential energy into radiation. Because their high luminosity attracts our attention, we tend to detect the brightest QSOs—those that have the largest black holes or the most abundant supply of fuel. (QSOs may include features of a normal galaxy, such as stars distributed in spiral arms, but these cannot be seen because of the distances.) Beyond a certain distance from a black hole, the gravitational field is no larger than that of a collection of stars of the same total mass. This means that stars or gas in the immediate vicinity of a black hole may be destined to accrete upon it and eventually become part of it, but matter beyond that distance is safe from accretion. Thus, the supply of fuel for the black hole is finite, and Seyfert galaxies probably

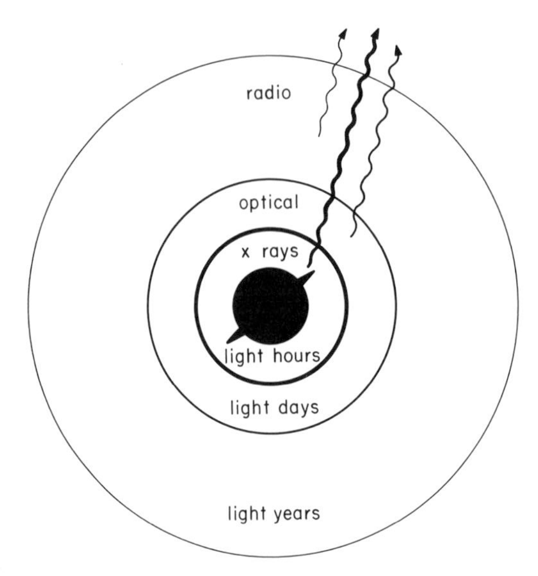

Figure 8.9
Sizes of the emission regions for x rays, visible light, and radio frequencies around a supermassive black hole, expressed in light travel time. The black hole is assumed to be at the center of a galaxy, accreting matter and converting gravitational potential energy into radiation through a complex series of processes. (Smithsonian Astrophysical Observatory illustration by John Hamwey)

represent a later phase than the QSO—a stage in which the supply of matter, and hence the accretion rate, has decreased. However, the fuel supply is still sufficiently large for the luminosity to be appreciable. Normal spiral galaxies, such as M31 and the Milky Way, are examples of a different condition. The supply of matter is virtually exhausted, or perhaps was not large to begin with, and the center of the galaxy is relatively but not completely quiet. Also, the central black holes in the Milky Way or other normal galaxies may be considerably smaller than those within the bright QSOs that attract our attention.

At first, a central black hole with a mass equivalent to a billion or more stars seems very strange. However, in many respects it is more natural than either the stellar-size black hole or the neutron star, two objects whose existence we accept easily from the detection of compact galactic x-ray sources and pulsars. The stellar-size black hole and the neutron star are formed explosively as the end products of massive stars. A massive black hole may be formed more gradually. According to the laws of physics, if 5 billion solar masses of stars or gas are concentrated to a density not much above that of an ordinary gas (a few milligrams per cubic centimeter) the ensemble will inevitably collapse to form a black hole. It is not too difficult to imagine this process occurring in the nuclei of these very young galaxies we detect as QSOs.

Clusters of galaxies

The large central object that dominates figure 8.2, the cluster of galaxies known as A2256, is a typical diffuse extragalactic source. X-ray sources that are extended in size, show spectra with temperatures above 20 million degrees, and are not coincident with a supernova remnant invariably turn out to be clusters of galaxies. Clusters of galaxies are associations ranging from a few hundred to a few thousand galaxies that form a closed dynamical system. Short of the universe itself, they are the largest aggregates whose dynamics can be studied in detail.

Study of the velocities of the galaxies within the cluster shows that they form an association in which the members are mutually bound by their own gravity. However, there is a severe problem: The masses of the individual galaxies seem to be insufficiently large by at least a factor of 10 to provide the gravitational binding energy needed to maintain the integrity of the cluster. The masses of the galaxies are determined by basing estimates of their mass-to-light ratio on that of stars within our own galaxy. The average star within our own galaxy has a mass-to-light ratio not more than several times that of the sun. Even if we allow for the fact that the cluster glaxies have a higher proportion of relatively dark stars than our galaxy, the cluster mass obtained by measuring the light is still too small. The puzzle of how A2256 and other clusters of galaxies are bound despite the insufficient mass seen in visible light has existed for some 45 years and is known as the problem of the missing mass. One of the reasons x-ray measurements of clusters are so interesting is that they provide new information on this classic problem.

The tendency of galaxies to form clusters and perhaps even larger associations known as superclusters is one of the major driving forces of nature. It affects not only the overall structure of the universe but possibly the evolutionary history of the galaxies themselves as a consequence of their interaction with other cluster members. Cluster formation is a contemporary process; more clusters of galaxies exist now than did in the early universe. Thus, we know that the structure of the universe is still evolving. For these reasons, the ability to identify clusters of galaxies at various distances and hence at various stages in their evolution is crucial.

The Einstein Observatory was able to study only a very small fraction of the sky in fine detail; however, its x-ray images of clusters of galaxies

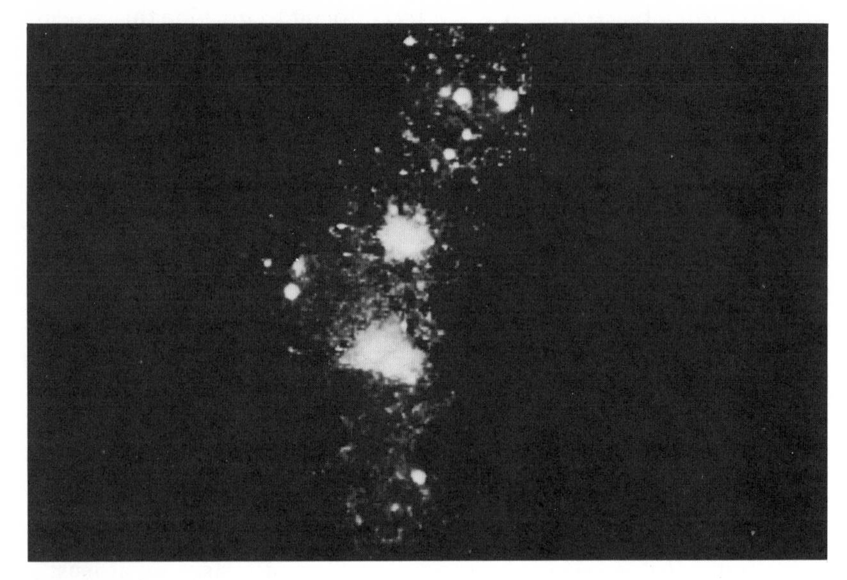

Figure 8.10
View of a region in the southern sky obtained with the Imaging Proportional Counter. Several 1° fields are joined together. The largest image (near the center) is that of two subclusters of galaxies probably in process of merging. The smaller image to the north is another cluster of galaxies that may be physically associated with the first. (Smithsonian Astrophysical Observatory photo)

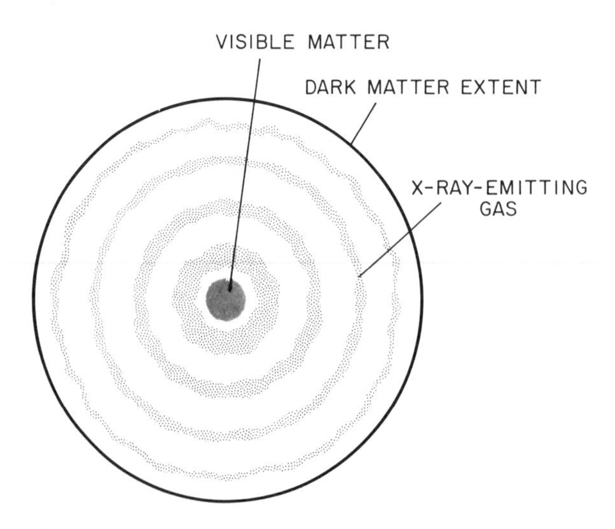

Figure 8.11
Distribution of mass in the giant elliptical galaxy M87. Matter can be seen in visible light only near the center. The hot, x-ray-emitting gas is found at much larger radii and is held in a gravitational field of a dark, massive halo. (Smithsonian Astrophysical Observatory illustration by John Hamwey)

reveal remarkable structures, as shown in figure 8.10. This region in the southern hemisphere contains several clusters of galaxies. It appears that two subclusters in the center of the picture are in the process of merging. Another cluster lies to their north. Of particular interest is the indication of a luminous region connecting all three clusters, a vast cloud of hot gas. Most of the background pointlike objects in this picture are, as usual, QSOs.

That gas with a temperature of 100 million degrees remains trapped in the medium between the members of a cluster of galaxies is highly significant. The mass of the gas we detect directly through its x-ray emission is not negligible. In fact, it is about equal to the mass we derive for the galaxies on the basis of mass-to-light ratios for normal stars. However, the mass of gas is insufficient by a factor of 10 to confine itself gravitationally. But when we calculate the amount of mass needed to establish a gravitational potential large enough to prevent the hot gas from escaping and to make it conform to the observed shape, it turns out that the mass needed to trap the gas gravitationally is equal to the same value of missing mass of the cluster required to explain the distribution of velocities. Ultimately, the x-ray data can do more than simply confirm the existence of the missing mass, although that is already highly significant. By measuring the temperature and density distributions of the hot gas, we can calculate the exact shape of the gravitational potential and hence trace out the contours of the missing mass. Perhaps in this way we may eventually identify it. We are still in the early stages of this work, but an important related result has already been obtained from the radio galaxy M87.

M87 was described above as an active galaxy probably containing a giant black hole at its center. It is also a member of the Virgo cluster of galaxies, situated at virtually zero velocity with respect to the average of the other cluster members and surrounded by a large halo of gas at a temperature of 30 million degrees. Moreover, M87 is situated at almost the optimal distance for studying the temperature and density distribution of this gas with the Einstein Observatory. By applying simple laws of physics to the analysis of the gaseous halo around M87, we can reconstruct the gravitational potential that contains the gas and determine the mass as a function of the distance from the center. A rather remarkable result is obtained. The stability of the gaseous halo requires that there be 10 times as much mass in M87 in unseen matter as there are in stars or gas. Furthermore, the dark matter follows the contours of the gas and extends considerably beyond the visible light (figure 8.11). If every galaxy member of the cluster A2256 were like M87 in having a dark halo 10 times as large as its visible mass, that would account for the missing mass. Of course, it is presumptuous at this stage to conclude that every galaxy has a massive dark halo. M87 could be atypical because of its unusual condition of being at rest in the Virgo cluster or because of some other peculiarity in its history, or the dark matter in clusters of galaxies could be distributed throughout the cluster and not attached to individual galaxies. Furthermore, the extra mass component could be a property of only members of rich clusters and not of galaxies in general.

Still, the results on M87 suggest two profound questions: What is the nature of the dark matter in the halo of M87? What would be the consequences if indeed every galaxy in the universe had a dark halo like M87's (more explicitly, 10 times as much mass as found from visible-light measurements)?

Because the x-ray-emitting gas of M87 extends out to a much larger distance than the visible matter, it is clear that the material that establishes the gravitational potential is quite dark—much darker than even the darkest ordinary stars. (The evidence for dark matter in other galaxies, namely the spiral types, is based on studies of the velocity of their rotation about their centers. Although these rotation studies cannot be carried out to as far from the centers of spiral galaxies as from the center of M87, and although they apply to a different kind of galaxy, they confirm the general picture that dark matter is present in abundance around galaxies.)

Three possibilities have been proposed to explain the dark matter. One is that it consists of a collection of neutron stars or black holes, the compact remnants of stars destroyed in cataclysmic supernova explosions. Another suggestion is that the halo is made of many objects not much larger than the large planets. These objects would be too small to generate the thermonuclear processes that produce visible light on the sun and other stars because their gravitational field is below the minimum amount needed for compression to a density sufficient to sustain nuclear burning. A third explanation has received the most attention recently. According to the widely accepted theories concerning the early history of the universe, an uncharged particle known as the neutrino should be omnipresent in copious amounts. Most theorists argue that the universe is in a virtual bath of neutrinos. Other varieties of particles may also be present, depending on the theoretical model one considers. The neutrinos and these possible other particles interact very weakly with ordinary matter and are thus very difficult to detect. Even when neutrinos are produced in the laboratory as highly concentrated beams, very large detectors and very long observation times are needed to detect them. The critical question concerns the mass of the neutrino. Neutrinos are so abundant that if their mass were only 1 electron-volt (2 millionths of the mass of the electron) they would dominate the mass of the universe. Whether neutrinos can account for the massive dark halo of M87 and the missing mass of clusters hinges, then, on whether the mass of neutrinos exceeds 10 eV. Laboratory experiments so far have been inconclusive, but the measurements believed to be the most precise do not support that large a mass. However, if the neutrino mass should fall short of explaining M87's halo, attention might focus on two other particles proposed by theoreticians: the gravitino and the photino. Less abundant than neutrinos, both would need masses exceeding 100 eV to explain M87's halo. Unfortunately, unlike neutrinos, they cannot be produced in the laboratory, so it will be virtually impossible to measure their masses. As the theory is still in its formative stages, it would not be surprising if other varieties of uncharged particles were to be proposed as the source of the dark halos. No matter what is the winning candidate for dark matter, it is becoming increasingly clear that most of the matter within clusters of galaxies and in the halos of large elliptical galaxies like M87 is in a form that emits no visible light.

There are profound implications in carrying this theory a step farther and assuming that every galaxy in the universe, be it a cluster member or not, has a dark halo 10 times more massive than the totality of its visible stars. This assumption cannot be justified at present. However, if it were true, it would have important consequences for the ultimate fate of the universe. A key question of cosmology is whether the expansion of the universe initiated by the "big bang" will go on forever or whether the outward flow of the galaxies will be halted and eventually reversed by mutual gravitational attraction. If the mass of the universe is sufficiently large, then it is destined to contract and reenact in reverse the great explosion in which it was created. Such a universe is said to be "closed," in contrast to an "open" universe, which expands indefinitely. So far the mass calculated from visible-light measurements is not much more than 10 percent of the mass needed to close the universe. On this basis, the evidence favors an open universe. But should it turn out that every galaxy is surrounded by a dark halo increasing its mass by some 10 times, astronomers would have to modify their views to favor a closed universe. A closed universe is a precursor to the ultimate black hole. If all the galaxies are fated to crash together, the residue of the explosion could be an entity with a gravitational field so intense that nothing escapes past the pulse of radiation marking that final event.

Conclusion

With the expiration of the Einstein Observatory, American x-ray astronomy entered a hiatus. In the immediate future, all new x-ray astronomy programs will be centered in Europe and Japan. Perhaps, after a period of consolidation during which new space-flight systems will become available, the United States will again be able to participate directly in this rapidly developing, exciting field. There is certainly no shortage of new concepts and bold ideas for larger and more effective x-ray telescopes with which to study the fundamental questions of cosmology.

9

The Future of Space Astronomy

George B. Field

Astronomy, like other sciences, develops in stages. First, new types of celestial objects are discovered when new instruments and new observing techniques are introduced. These objects are classified, and similarities and differences are noted. Then, as more powerful telescopes are trained on them, their properties are analyzed in detail and theoretical models are proposed to account for their apparent behavior. Finally, predictions of new properties based on the theoretical models are tested against new observations. If the models pass all observational tests, they are generally accepted as representing our best understanding of the objects, and astronomers turn to the study of less-understood objects.

Advantages of space astronomy

Why is space astronomy important in this process? Previous chapters in this book should have already provided answers to this question. Certain objects emit electromagnetic radiation most copiously in wavelengths at which Earth's atmosphere is opaque. Thus, we would be largely ignorant of the violent processes occurring in the outer envelope of the sun, of the properties of neutron stars and black holes, of high-speed streams being ejected from stars, and of dense clouds condensing to form new stars were it not for the ultraviolet, x-ray, and infrared observations carried out from platforms high in or above our atmosphere. Of course, there also would have been no way to explore the magnetospheres, atmospheres, and surfaces of the planets and their satellites without space probes. However, I will not discuss such probes further, but will focus on space astronomy, which I define as remote observation of astronomical objects from Earth orbit above the atmosphere.

There is another aspect of the space environment that has only just begun to be exploited for astronomy. Not only does the Earth's atmosphere block out radiation of certain wavelengths completely; it also severely distorts the optical, or visual, images of those objects we can see through the atmosphere. Everyone is familiar with the "twinkling" of the stars; they appear to dance about like fleas because refraction of their rays by moving pockets of warm air constantly changes their apparent positions. This effect, called "astronomical seeing," limits the ability of all ground-based telescopes, even the largest, to produce images of small diameter.

The quality of seeing can be improved by placing telescopes on mountains above most of the atmospheric turbulence, but even on the highest peak the best images will not be much sharper than 1 arc-second—the angle subtended by a dime at a distance of a mile. The diffuse background light surrounding an observed object, which is equivalent to one 23rd-magnitude star in every square arc-second of sky, also makes it difficult to detect very faint stars. By contrast, viewing objects from above the atmosphere gives images whose sharpness is limited only by the optics of the telescope—typically, to a tenth or a twentieth of an arc-second in diameter. Since the background light in neighboring sky regions of this size is equivalent to only one 29th-magnitude star, a space telescope can, in principle, observe much fainter stars.

A third significant advantage to observing from space is that, even for those wavelengths that can be detected on Earth, such as the optical band, the atmosphere is never perfectly transparent. The transparency varies with direction in the sky, and it is often diminished or eliminated by local weather. Moreover, at certain wavelengths, molecules in the upper atmosphere emit radiation that is observable by ground-based telescopes, thus confusing the observations of stars. And, increasingly these days, radiation from street lamps, neon signs, and broadcasting can interfere with observations of certain wavelengths. The ability to avoid all these problems makes space astronomy even more attractive for the future.

Indeed, for the astronomer, space technology may provide the only means for answering the ultimate question: What is the universe? Our present understanding is based on an astonishingly simple idea: that the universe, born in a "big bang" some 10–20 billion years ago, has expanded smoothly since then. Some billion years after the "big bang," matter began to condense into galaxies, and then into stars (of which our sun is rather typical). However, even if this simple scenario is correct, it raises many questions about how and when and why these events occurred.

Questions for the future

At any one time, modern astronomy is dealing with several broad questions, about which we know enough to state the problem but not enough to give convincing answers. A 1982 report by the National Academy of Science's Astronomy Survey Committee, of which I was chairman, identified seven such questions:

- What is the large-scale structure of the universe?
- How do galaxies evolve?
- What role do violent events play in the evolution of the universe?
- How are stars and planets formed?
- What causes activity on the surfaces of the sun and other stars?
- How widespread are life and intelligence in the universe?
- Do the connections between astronomy and the fundamental forces of nature hold the key to a unified understanding of all cosmic processes?

Although these questions may be impossible to ever answer definitively, significant progress on them can be made in the decade ahead. In the remainder of this chapter I will discuss how specific programs and instruments proposed by the Astronomy Survey Committee, particularly in space astronomy, will address them in the 1980s and the 1990s.

The Space Telescope and ground-based astronomy

After nearly a quarter-century of planning and development, the National Aeronautics and Space Administration in 1985 will use its Space Transportation System, the "Space Shuttle," to launch the Space Telescope. A 2.4-meter reflecting telescope usable at wavelengths ranging from the mid-ultraviolet (1,200 Ångstroms) to the near-infrared (12,000 Ångstroms), the Space Telescope is planned as a permanent observatory for ultraviolet, optical, and infrared observations. The curvature of its primary mirror will be precise enough to obtain stellar images only 0.05 arc-second in diameter (a value determined by a fundamental diffraction limit associated with its diameter). This will permit images of the outer planets as sharp as those made of Jupiter by the Voyager

Figure 9.1
Artist's conception of the Space Telescope. (NASA illustration)

spacecraft at its closest approach and images of galaxies 20 times sharper than those made by telescopes on the ground.

The Space Telescope will really come into its own by imaging galaxies. Ground-based telescopes now see the most distant galaxies merely as points of light; the Space Telescope will resolve them and make it possible for them to be classified as standard Hubble types. But will the distant galaxies fit these classical molds? Have galaxies evolved into different shapes during the time it takes their light to reach us from the regions of space so remote that this time is a significant fraction of the age of the universe? Current theories predict significant evolutionary changes; if these changes are not found, we may have to question our model of the universe.

The Space Telescope will be able to observe objects 10–100 times fainter than those now seen, making it possible to detect and study galaxies at distances near the limits of the observable universe (where, because of their high luminosities, we have so far been able to detect only quasars). A crucial question in cosmology is whether galaxies at remote distances—that is, those seen as they were when the universe was only a fraction of its present age—are arranged into groups and clusters, as are those located much closer to us. According to one theory, galaxies began to swarm together when they were mere infants;

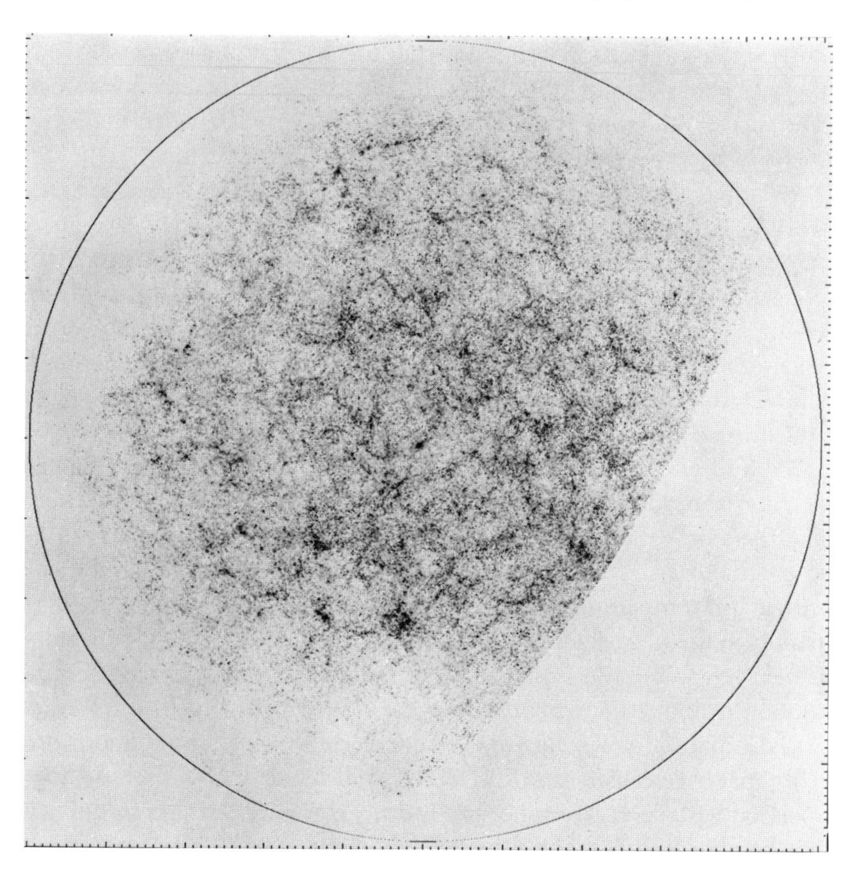

Figure 9.2
Computer-produced image showing the positions and luminosities of the million brightest galaxies in the northern hemisphere of the sky. Note that they are not distributed at random but are clustered by their mutual gravitational attraction. Each dot represents from one to ten galaxies; the lightest dots representing one galaxy and the darkest ten. (Courtesy of P. J. E. Peebles, Princeton University)

however, another theory suggests that the gravitational forces drawing them together did not take effect until the galaxies reached adolescence. The Space Telescope should be able to distinguish between these two theories by observing the clustering of distant galaxies.

A 1-meter telescope called the Far-Ultraviolet Spectrograph in Space (FUVS), to be launched later, will extend the observable range of ultraviolet wavelengths down to 912 Ångstroms (beyond which the opacity of interstellar gas makes observations very difficult). This is an important improvement because many of the most abundant atoms and molecules in astronomical objects absorb and emit radiation at characteristic wavelengths in the region between 912 and 1,200 Ångstroms.

To address the completely different problem of solar and stellar activity, the Space Telescope and the FUVS will be turned to stars like the sun, which are densely distributed near the solar system. We know that the sun is surrounded by an envelope of superheated gases, ranging in temperature from 50,000°K to 2,000,000°K, which radiates mainly in the ultraviolet and x-ray parts of the spectrum. This envelope is believed to be the consequence of magnetic fields generated by the motion of ionized gas deep inside the sun, which float to the surface and dissipate, thereby heating the envelope to 1,000,000°K and more. The Space Telescope and the FUVS will detect spectral features associated with such envelopes around nearby stars. Although they will not be able to image the envelopes of these stars, the telescopes will be able to check evidence from recent ground-based observations that these stars rotate slowly and that the magnetic activity believed to heat their envelopes waxes and wanes like that of the sun.

Despite the power of these new space observatories, there will still be a need for ground-based telescopes. For example, although the Space Telescope will detect very distant galaxies, its aperture is not large enough so that the light collected can be broken down into detailed spectral information. A superlarge optical telescope, with perhaps 30 times the collecting area of existing instruments, could obtain the spectra of galaxies found by the Space Telescope; this is the key to determining the distances and motions of the galaxies. In fact, a number of large ground-based optical-infrared telescopes are planned for use in conjunction with the Space Telescope and the FUVS. A revolution in telescope-building technology, initiated with the 1979 opening of the 4.5-meter Multiple-Mirror Telescope (MMT) of the Smithsonian Institution and the University of Arizona, will make possible the design and construction of much cheaper ground-based telescopes. The MMT cost only about a third as much as a telescope of equivalent

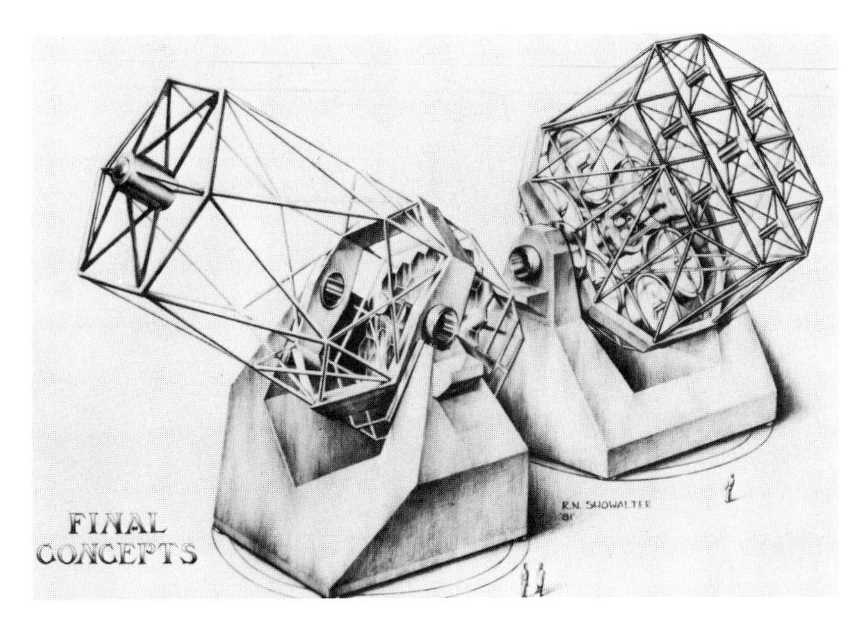

Figure 9.3
Two designs proposed for the 15-meter National New Technology Telescope. (Kitt Peak National Observatory illustration)

size built according to conventional designs. Because its six primary mirrors are unusually light, the weight of the mount and that of the optical support structure are reduced correspondingly. And because the six mirrors are essentially used as separate 1.8-meter telescopes rather than as one 4.5-meter mirror, the focal length is reduced; this, when combined with an altitude-azimuth mount, make possible a more compact and less expensive building.

The small community of telescope builders is convinced that further refinement of the principles embodied in the MMT should make it possible to build a telescope with a 15-meter aperture for about 10 times the cost of the MMT. The University of Texas is planning a telescope having a single 7-meter mirror; the University of California has proposed a 10-meter instrument based on a single surface constructed of 36 individual 1.8-meter segments, and the University of Arizona and the Smithsonian Institution have suggested a scaled-up version of the MMT. The proposed 15-meter National New Technology Telescope may be based on one of these designs.

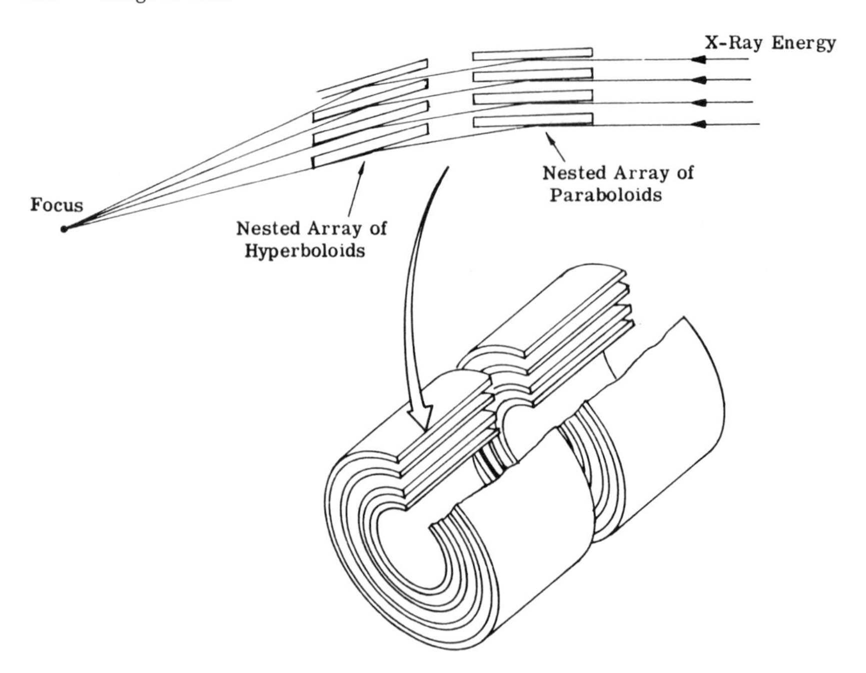

Figure 9.4
Schematic drawing of the grazing-incidence optics of the Einstein (HEAO-2) satellite. (Smithsonian Astrophysical Observatory photo)

The Advanced X-ray Astrophysics Facility

At the time of this writing, the United States does not have a single high-energy astrophysics instrument in orbit, even though American astronomers founded x-ray astronomy and made it a major contributor to astronomical knowledge. For this reason, the NAS Astronomy Survey Committee gave its highest priority for major new space programs in the 1980s to an Advanced X-ray Astrophysics Facility (AXAF), the dream of a generation of x-ray astronomers who discovered x-ray sources and used rockets and satellites to probe their nature in the 1960s and the 1970s.

Designed as a permanent observatory in Earth orbit, AXAF will be capable of making high-resolution images of cosmic x-ray sources with a sensitivity up to 100 times that of its predecessor, HEAO-2 (the Einstein Observatory). AXAF will be a reflecting telescope using grazing-incidence optics to form precise images in the x-ray band between 0.2 and about 8 keV (between 1.5 and 60 Ångstroms wavelength). The telescope will have an aperture of 1.2 meters (the Einstein Observatory's was 0.6 meter) and an angular resolution 10 times that of the Einstein Observatory. This means that AXAF may study supernova remnants in our own galaxy with 10 times the precision and will be able to observe coronal x rays from virtually all of the 100,000 stars nearest the sun, thus complementing in the x-ray region the ultraviolet studies of the envelopes of normal stars by the Space Telescope and the FUVS. Individual x-ray binary stars—systems in which a black hole may be swallowing gas emitted from a normal stellar companion—will be observable all the way out to the Virgo cluster of galaxies, at a distance of 60 million light years; this will increase tenfold the number of these objects available for study.

AXAF also will have sufficient sensitivity to probe deep into the realm of the quasars. Not only will it study quasars themselves (most of which are strong x-ray emitters), but it will also detect the hot (100 million °K) gas in clusters of galaxies at the limits of the universe. Both capabilities will contribute to a better understanding of the evolution of galaxies. For example, quasars are now believed to be an evolutionary stage of some galaxies—a stage in which black holes with 100 million or more solar masses are formed in the galactic nuclei. As the black hole swallows up the surrounding gas and stellar material, the gas is heated to 10 million °K or more. By observing variations in the x-ray emissions of a quasar within only a few hours, the Einstein Observatory showed that regions where this happens are very small—only a little larger than the solar system. Such a region may produce up to a million million times the luminosity of the sun.

In addition, AXAF may detect quasars many times more remote than the most distant so far observed. Such detection will test whether the apparent "edge" or limit to the realm of the quasars at a red shift of 3.5 (approximately 10–15 billion light years) is real or only an artifact of our present observing techniques, and whether most of the diffuse x-ray background is actually due to large numbers of still undetected quasars as most investigators believe. This is important because an alternative theory would attribute the diffuse x-ray background to a thin intergalactic gas pervading all space, having a total mass roughly equal to that of the galaxies, and heated to over a billion degrees by some unknown process.

AXAF's capabilities for studying x-ray emission from hot gas in clusters of galaxies will continue the investigations started by the Einstein Observatory and will extend much deeper into space. More detailed x-ray spectrographs will permit the detection of spectral features from highly ionized iron atoms, allowing determination of the temperature of the gas, its iron abundance, and even its red shift. These data are important for understanding the early evolution of galaxies—how and why they first clustered and what processes stripped the gas out of them to form the intracluster medium. Theories based on an evolving universe suggest that the x-ray emissions of distant clusters should differ from those of nearby ones; AXAF will test this prediction.

AXAF will be complemented in Earth orbit by the Gamma-Ray Observatory (GRO), the Extreme Ultraviolet Explorer (EUVE), and the X-ray Timing Explorer (XTE), each looking at a different wavelength band in the realm of high-energy photons. The GRO will map the universe in gamma rays, studying sources in the 1–100-mega-electron-volt region, including those first discovered by the European COS-B satellite. The small EUVE will carry out the first exploration of the whole sky for sources emitting in the 100–912-Ångstrom band. The XTE will study x rays emitted by variable sources, carefully monitoring their changes in search of periodic fluctuations.

The GRO, already funded and scheduled for launch in 1987 or 1988, will be sensitive to explosive phenomena in the universe, including a puzzling source of radiation at the center of the Milky Way. There is also gamma-ray evidence that positrons (antielectrons) exist in large numbers at the center of the Milky Way, but it is not known how these were created. The GRO will also observe the spectral features in the gamma-ray region that characterize freshly synthesized heavy elements. If, as is believed, all the heavy elements in the universe (including those composing the Earth) were created in supernova explosions, the GRO should be able to obtain direct proof of this by observing these features in remnants of recent supernova explosions.

The Very-Long-Baseline Array

The Survey Committee's second priority among major new programs for the 1980s went to the construction of a Very-Long-Baseline Array (VLBA) of radio telescopes that will provide extremely sharp images at radio wavelengths (0.0003 arc-second resolution). Although primarily a ground-based project, the VLBA will be complemented by a single antenna to be launched into Earth orbit. This will allow astronomers to generate radio images three times sharper than those obtained with the ground-based array alone.

Two key astronomical problems are already being attacked by the coordinated use of various radio antennas in an informal array. The first is that many quasars emit jets of matter in well-defined directions, interpreted by some theorists as being perpendicular to a disk of ultrahot gas orbiting a massive black hole. In some cases, the blobs of matter appear to be moving away from the parent quasar at speeds exceeding that of light by much as a factor of 10. This can be understood with respect to Special Relativity if the blobs are traveling at nearly the speed of light and nearly toward the observer. (Unfortunately, such an interpretation does not explain why the jets just happen to be directed at us.) The VLBA will also be sensitive to interstellar masers, phenomena that occur in very cold interstellar gas clouds. Radio astronomers have found sharp peaks of radiation from water molecules and hydroxyl radicals, apparently originating in superdense solar-system-size gas clouds where the gas is excited to emit in the manner of a laboratory maser or laser. Some of these objects may be new stellar systems in the process of formation. Whatever their nature, their positions will be determined by the VLBA to within a few ten-thousandths of an arc-second (the angle subtended by a bacterium at a distance of a mile). Moreover, by tracking their motions and using some elementary geometry, we will be able to determine their distances precisely even if these objects are located on the other side of our galaxy. This is a totally new and independent way of determining astronomical distances within our galaxy. The first application of this technique gave a distance to the Orion Nebula that agreed with distances determined by previous methods to about 10 percent. Particularly exciting is the VLBA's potential for determining the distance to the Andromeda Nebula by observing directly its rotation, known from the Doppler shift in its spectrum. Previous estimates of the distance to Andromeda are indirect and somewhat unreliable. Since our knowledge of the size of the observable universe depends on an accurate knowledge of distances both inside and beyond our galaxy, the VLBA is expected to make a major contribution to cosmology.

In a proposal related to radio astronomy, the NAS Astronomy Survey Committee recommended that a search for extraterrestrial intelligence (SETI) be undertaken at a modest level and on the basis of competitive peer review of proposals. Naturally, such a search would be both difficult and chancy; humanity may well be alone in the universe, or at least alone in being able to communicate in symbols. Nevertheless, many astronomers believe that intelligent life, like the phenomenon of life itself, is as much a part of the astronomical universe as stars and galaxies. Even beyond its profound philosophical implications, the discovery of even one intelligent civilization could yield information of scientific value, such as astronomical data on the planetary system and environmental and biological data on the habitat. The best way to approach this problem is not known. However, from an analysis

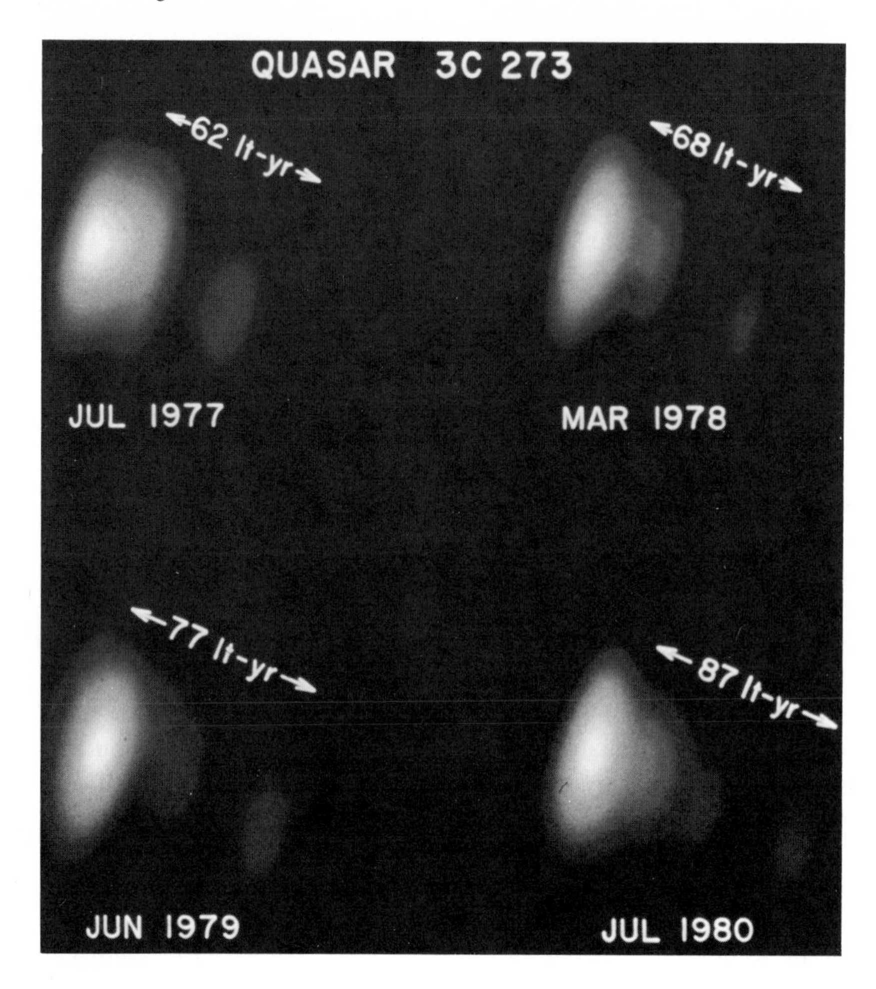

Figure 9.5
Blobs of relativistic electrons leaving the quasar 3C273 at velocities that appear to exceed that of light. (Courtesy of Owens Valley Radio Observatory, California Institute of Technology)

of energy requirements, ease of reception, speed of communication, and other factors, many scientists have concluded that it makes sense to begin the search for intelligent signals in the radio-frequency band. Most efforts so far have taken this approach. However, it is possible that someone will demonstrate overwhelming advantages to other means of communication; in that case, we should be prepared to pursue other approaches. In the meantime, it is worth noting that the VLBA will constitute a very large collecting area at very short radio wavelengths, should such wavelengths prove of interest for SETI. Ultimately, it may prove desirable to establish a large radio telescope (say, one with a 300-meter aperture) in space, where SETI could be conducted without man-made radio interference.

A large deployable reflector in space

Fourth in priority among the major new programs urged by the NAS committee was a Large Deployable Reflector (LDR) to be launched into Earth orbit for observations in the far-infrared and submillimeter wave parts of the electromagnetic spectrum. The facility is planned as a 10-meter reflector, opening to full operational size only after it is removed from the Space Shuttle's bay. The LDR will be preceded into space by a number of other instruments operating in this wavelength range, including the Shuttle Infrared Telescope Facility (SIRTF) and the Cosmic Background Explorer (COBE). The SIRTF will be only about a meter in diameter, but its optics, unlike those of LDR, will be cooled to extremely low temperatures to eliminate background interference and to increase sensitivity. The COBE, as its name implies, is designed specifically to study the cosmic microwave background radiation thought to have originated in the "big bang." Its ability to measure spectra in the millimeter and submillimeter parts of the spectrum with a precision of about 0.01 percent depends on its being above the atmosphere, which absorbs and radiates energy copiously at these wavelengths. Balloon-borne experiments have suggested that the background spectrum deviates significantly from its theoretically expected thermal form; if this is true, the COBE will observe the discrepancy and determine its dependence on wavelength—a key datum for cosmology.

Objects that radiate mostly in the infrared and submillimeter parts of the spectrum have temperatures between 10°K and 1,000°K. This includes the gas and dust in dense interstellar clouds where stars are forming; possible protoplanetary disks, such as the one believed to have given birth to our own planetary system; planets, comets, and

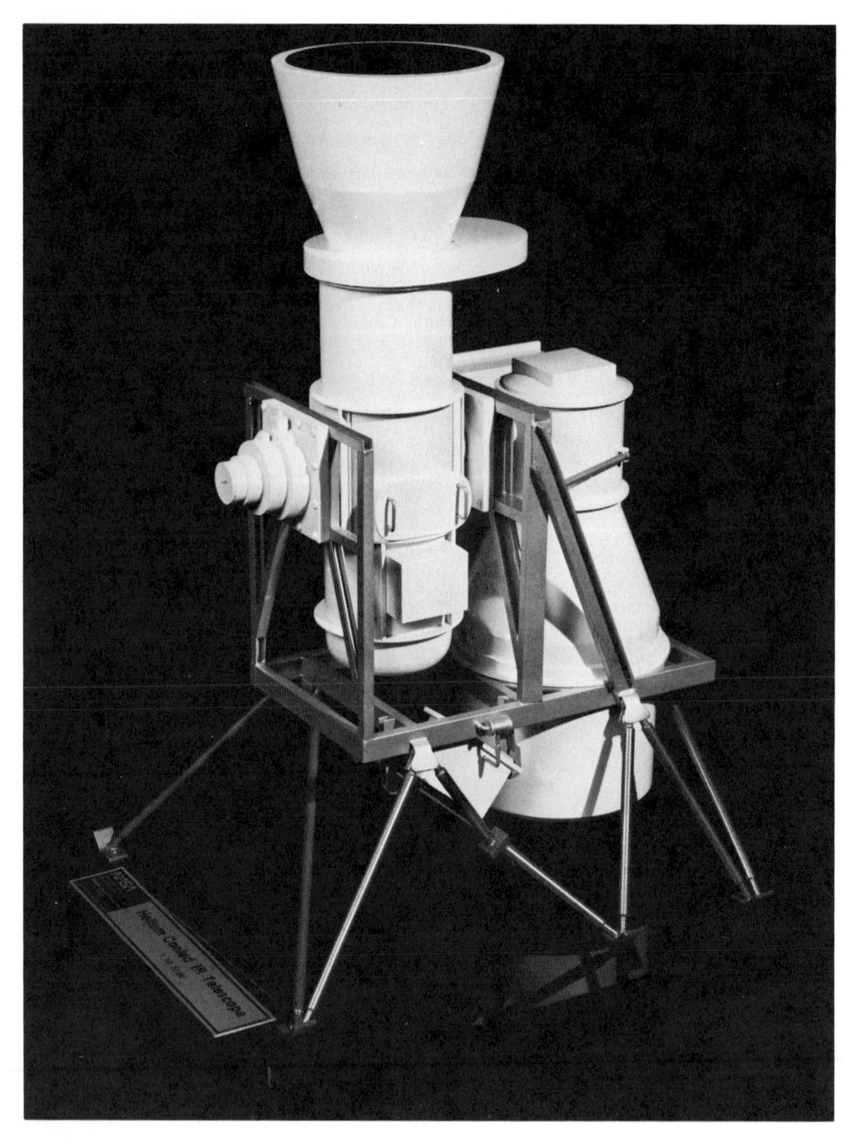

Figure 9.6
Model of the Shuttle Infrared Telescope Facility. (NASA/Marshall Space Flight Center photo)

asteroids; and cool stars, particularly those ejecting gas and dust into space. The SIRTF should discover thousands of such objects and obtain the spectra of the brightest. The SIRTF's mirror is too small to obtain detailed spectra of fainter objects, but the LDR, with 100 times the area, will do so. The LDR also will have the advantage of higher angular resolution. Telescopes operating at the relatively long wavelengths in this spectral region are limited in resolution by diffraction effects; however, the larger the aperture, the higher the resolution. Thus, at a typical operating wavelength of 2.5 microns, the SIRTF will have a resolution of 5 arc-seconds—sufficient for discriminating objects from their neighbors, but inadequate for forming precise images of extended objects. In comparison, the LDR will have a resolution of 0.5 arc-second, and will thus be able to penetrate the cores of dark clouds where stars are forming and to map in detail these processes of stellar incubation. Indeed, at an operating wavelength of 25 microns, the angular resolution of 0.5 arc-second is sufficent to resolve preplanetary disks anywhere in the dark, star-forming clouds of Ophiuchus or Taurus at distances up to 600 light years. Moreover, the LDR's spectroscopic capability will enable us to study a wide variety of atoms and molecules in such regions and to determine the temperatures, densities, and velocities of the gases, which are keys to understanding the gas flows that are crucial to the creation of stars.

Solar physics

Many other planned and proposed projects using the unique advantages of space platforms will affect astronomy profoundly. One area in which significant advances are expected is solar physics. As I have already discussed, the hot gaseous outer layers of the sun are best studied in the ultraviolet and x-ray bands. Recent discoveries indicate that these layers are far more complex than was thought even 10 years ago. Solar magnetic fields are apparently created and transported by complex motions of gases below the surface in a process involving both vertical convection and zonal winds much like those in the Earth's atmosphere. Once they emerge above the surface these fields bend and snap like taut springs, releasing energy in several different ways, some directly affecting the Earth: the solar wind, which energizes the Earth's radiation belts; the solar flares, which disrupt radio communications; and the formation of magnetic regions, which determine the amount of sunlight reaching the Earth and, hence, the temperature of the Earth.

Figure 9.7
The Orion nebula, a star-forming region. (Smithsonian Astrophysical Observatory photo)

By the late 1980s, solar astronomers plan to launch aboard the Shuttle a Solar Optical Telescope (SOT), a precise 1-meter optical-ultraviolet telescope with an aperture sufficient to obtain detailed spectra of solar gases. It will have a resolution of 0.03 arc-second in the far-ultraviolet range, adequate to resolve features as small as 20 kilometers on the sun. The high intensity of solar radiation will allow measurements to be made rather rapidly, and the SOT will be operated from the Shuttle's bay for a week or two rather than injected into orbit. On subsequent flights of the SOT, the telescope will be coupled with ancillary instruments working at extreme-ultraviolet, x-ray, and gamma-ray wavelengths. These will be mounted alongside the SOT to make simultaneous observations of the active regions on the sun. Ultimately, the entire complement of instruments is to be injected into orbit as a long-lived, free-flying mission known as the Advanced Solar Observatory. This collection of instruments will constantly probe the activity of the sun's outer layers. Because the sun is the only star we can study in detail, these observations will become benchmarks

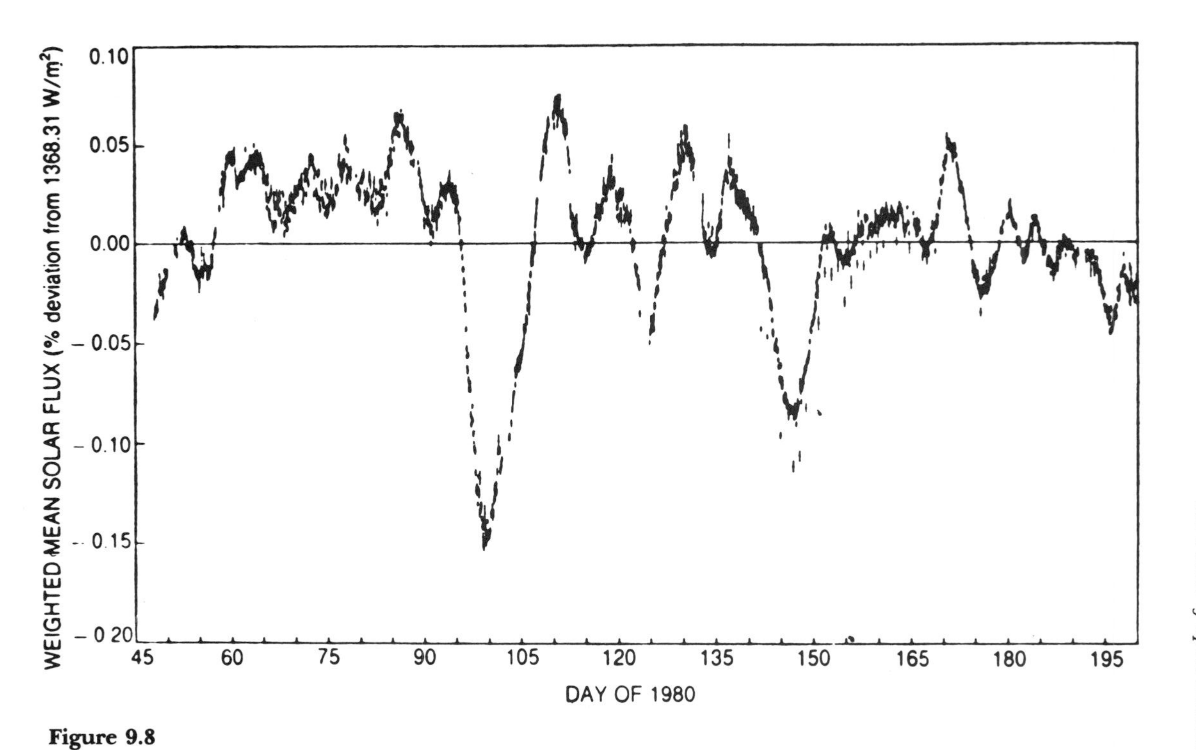

Figure 9.8
Total solar luminosity at Earth's orbit shows fluctuations that correlate with the passage of sunspots across the visible disk. The solar flux, measured by a cavity radiometer on the Solar Maximum Mission, is plotted here as a percentage variation about the weighted mean for the first 153 days of the mission. Individual points represent the mean irradiance for the sunlit portion of one orbit. The large decreases near days 100 and 145 are concomitant with the passage of large sunspot areas across the solar disk. (NASA/Jet Propulsion Laboratory/Caltech illustration)

for the study of activity in a number of solar-type stars by the AXAF, the Space Telescope, and the FUVS.

The 1990s

The Astronomy Survey Committee also recommended a number of programs for study and development during this decade for implementation in the 1990s. From a much longer list, I have selected two for discussion here.

A very large space telescope for optical, ultraviolet, and infrared observations would be an extraordinarily powerful instrument. With a diffraction-limited aperture of 30 meters, it could observe over the whole wavelength range from 1,000 to 100,000 Ångstroms. Because of its great collecting area, this instrument could study objects 100 times fainter than those the Space Telescope can observe. One could select for detailed study any one of the 100 billion stars in the Andromeda nebula. The great angular resolution (0.004 arc-second) attainable with such an instrument would enable us to image directly features as small as half a light year within the nuclei of nearby active galaxies, and perhaps to see phenomena associated with massive black holes if they are present.

The committee also proposed the development of optical and infrared space interferometers. Such instruments, like the ground-based VLBA in the radio band, would overcome the limitation on angular resolution that results from the concentration of a telescope's collecting area in a single filled aperture. Interferometry is a technique for "opening" this collecting area by the placement of multiple apertures (telescopes) at widely separated points. The expanded collection area thus can take advantage of the fact that small changes in the direction of wavefronts arriving from astronomical objects result in measurable phase differences. These phase differences can be detected as shifts in the pattern of positive and negative interference between the waves observed at the focus of an instrument. Because the separation between neighboring areas of positive interference is proportional to the wavelength of the radiation, and inversely proportional to the physical separation of the telescopes in the interferometer, the separations required to achieve a given angular resolution are proportional to the wavelength. Thus, at centimeter wavelengths in the radio region the interferometer elements must be separated by continent-spanning distances of 2,000 kilometers to achieve a resolution of 0.001 arc-second, but at the shorter optical wavelengths the same resolution can be achieved with separations of only 100 meters.

One may ask why astronomers have not already built optical interferometers with such separations. They have, but installations such as the one at Narrabi, Australia, are limited to observations of only the very brightest stars because of atmospheric seeing effects. The constantly changing refraction along the wavefront paths through the Earth's atmosphere causes the interference pattern at the focus of a ground-based optical interferometer to vary erratically. Above the atmosphere, however, seeing conditions should be completely calm, and it should be possible to exploit the principles of interferometry at optical and infrared wavelengths in a straightforward manner. There are now proposals for optical interferometers in space with elements separated by rigid structures some 10 meters across. Farther in the future are interferometers in which the elements are independent, free-flying spacecraft 10 kilometers apart. The angular resolutions of these devices would be 0.01 and 0.00001 arc-second, respectively. The latter device would be able to resolve objects as small as two light weeks across within the nearest quasars, thus penetrating the very cores of these enigmatic objects.

Conclusion

In many ways, space is the only natural place to carry out astronomical observations. Observatories above the atmosphere can operate at virtually any wavelength. Furthermore, the absence of atmospheric effects makes possible great improvements in angular resolution, with the only limitation the precision of the instrument itself. Understandably, perhaps, some people conclude that advanced space technology will make it possible to carry out all astronomical research, including optical and radio astronomy, from space. Although this is true in principle, I do not think it is likely to happen in practice for some time to come. The reason is simple: Even with the enormous payload capacity of the Space Shuttle (30 tons), it is still expensive (about $500 per pound) to launch instruments into orbit. It is even more expensive to design, construct, and test astronomical instruments so that they will operate trouble-free for long periods in space.

Thus, in my opinion, for the immediate future, all the astronomy that can be done from the Earth's surface will continue to be carried out there, and space astronomy will be limited to those observations that can be carried out only from space. In the decades ahead, both ground-based and spacecraft instrumentation will be applied as appropriate to explore the limits of our astronomical knowledge. These limits involve both the large-scale structure of the universe (including regions of space so distant that the observed radiation began its journey to Earth not long after the universe was created) and the fine details of astrophysical processes much closer to home (such as those in the outer layers of the sun, which equally challenge our understanding of how the universe works).

10

Epilogue: The Rediscovery of Earth

Ursula B. Marvin

Sputnik I was a stunning surprise to people around the world, but the news was especially traumatic for Americans—including some of us who would not have expected to experience any special reaction. My husband and I were working in Brazil at the time of its launch, exploring remote areas of jungle and highland for ore deposits. We were deeply engrossed in the landscapes, the people, and our geological fieldwork, and were only dimly aware of the impending space age. Nevertheless, the moment I heard of Sputnik, I was swept by a powerful sense of shock and dismay—and astonished at my own response. I knew instantly that something momentous had occurred, that science and technology had taken a quantum leap, and that the United States' participation was less than I wished.

My concern continued, off and on, for weeks. I felt no better when the October issue of *Scientific American* finally caught up with us, bearing the following message in its "Science and the Citizen" column:

> . . . the satellite projects were not going too well. Scientists of the U.S.S.R. had not yet made laboratory models of their satellites or even decided on their size or weight. In the U.S., workers on Project Vanguard have built 20-pound models, but the propulsion problem is still so formidable that they think they may have to begin with projectiles no bigger than a softball, carrying no instruments except possibly a radio transmitter for tracking purposes.

By the time we read that, Sputnik II, weighing over 500 kilograms, was already in orbit, carrying several kinds of instruments and a dog.

I wish I could report that during my own first dark days of the space age I secretly vowed that I would someday devote my best efforts to space science. I had no such thoughts, but that is the way it worked out anyway. Within a couple of years after Sputnik I we left Brazil and began looking for career possibilities at home. By the great good fortune of happening to be at the right place at the right time I was offered an opportunity to join a research effort on the mineralogy of meteorites at the Smithsonian Astrophysical Observatory. Eventually, my interests—and those of the observatory—extended to the study of lunar samples and grains of interplanetary dust, and then branched sideways into the possibilities of measuring the Earth's crustal motions to test the theory of continental drift.

As a geologist, I had focused all my attention on the rocks and minerals of the Earth's crust. It did not occur to me then that we might learn more about Earth by studying samples of other bodies in the solar system or by examining the Earth, the moon, and the planets from the vantage point of space. However, this has been the course of Earth science during the past quarter-century. We have abandoned our view of Earth as an isolated body orbiting the sun through empty space. Earth science has merged with planetary science, and we have learned to think of Earth in the context of the solar system. This change in viewpoint and the fresh insights it generated brought about a second revolution in the Earth sciences. The first revolution involved the overthrow of the stabilist theories that had ruled geology since the nineteenth century and replaced them with plate tectonics, a mobilist theory that envisions large-scale horizontal motions of rigid crustal plates. Because the plate-tectonics revolution took place during roughly the same years, it sometimes stole the limelight from the planetary-sciences revolution. However, I believe that in the long run we will find these two revolutions to be of equal importance to our understanding of the planet we live on.

My intent here is to review some of the things we have learned about Earth by applying the techniques of space science and technology. Among a host of possible choices, I will focus on our new knowledge of Earth's radiation belts, gravity field, atmosphere, and surface features. Moreover, by describing the new information on Earth's age and early history provided by the study of meteorites and the lunar surface, I hope to offer a traditional "field geologist's" perspective on the theme of comparative planetology discussed above by James Head and John Wood.

Satellites and trackers

Artificial satellites did not, of course, spring into orbit spontaneously. Sputnik should not have come as a surprise to scientists or engineers.

Decades of experimentation with rockets had led to designs for orbiting satellites. Launches were intended by both the United States and the Soviet Union during the International Geophysical Year, which extended from July 1, 1957, to December 31, 1958. Committees of the IGY had organized a worldwide system for tracking satellites optically and by radio. More specifically, the Soviets had issued several statements that they were developing satellites, and in June 1957 they had officially notified IGY headquarters that their program was ready. Practically nobody took them seriously. The power of their rockets was cloaked in military secrecy.

October 4, 1957, was a Friday. In Washington that evening, IGY representatives who had spent the week discussing plans for the orbiting of satellites were attending a party at the Soviet Embassy. Presently, the chief American representative was called to the telephone. He returned with an announcement: "Gentlemen, our Russian colleagues are to be congratulated. They have successfully launched an Earth satellite."

One American delegate who had not stayed for the party was Fred L. Whipple, Director of the Smithsonian Astrophysical Observatory (SAO) in Cambridge, Massachusetts. The news broke just as his plane was approaching Logan Airport in Boston, but he did not hear of it until he arrived home. He turned around at the front door and rushed to the observatory. As part of its participation in the IGY, the SAO had accepted responsibility for the optical tracking of satellites. It had started building a worldwide network of twelve observing stations, but none were yet in operation. It had also organized Project Moonwatch, an international volunteer effort in which dozens of teams of enthusiastic amateur astronomers would systematically scan the night skies for satellites. Most teams had been trained and stood ready to observe, but none expected a launch for at least another six months. How would it be possible to alert some 2,000 Moonwatch volunteers that their duties should begin immediately? Very shortly, SAO staff members came crowding back to the observatory; soon they were followed by television, radio, and newspaper reporters. Dr. Whipple arrived and stayed until 4 A.M. while telephone and cable messages were sent around the world. By morning, more than 100 Moonwatch teams were preparing to chart the course of Sputnik I.

The Soviets furnished no orbital data for several days. However, by then a fairly good track of Sputnik I had been established from Moonwatch observations and from military tracking data. Two things that became apparent very early were that the resistance of the atmosphere was appreciable even at an altitude of 240 kilometers, as

Figure 10.1
Philippine Boy Scouts preparing to scan the sky for artificial satellites as part of the international network of volunteer observers "Moonwatch" organized by the Smithsonian Astrophysical Observatory. (Smithsonian Astrophysical Observatory photo)

was shown by a slow spiraling of the satellite toward Earth almost immediately after it was launched, and that the shape of Earth's gravity field would have to be recalculated because the satellite did not drift westward as rapidly as had been predicted from Earth-based measurements. More precise information required data from subsequent satellites, but two major preconceptions about Earth had been shattered by the very first spacecraft.

The Van Allen radiation belts

Explorer I, the first U.S. satellite, launched on January 31, 1958, carried equipment designed by James Van Allen of the University of Iowa to measure cosmic radiation. At first its Geiger counters registered radiation levels that increased with altitude, as was expected. When Explorer I rose above 800 kilometers, however, the counts rose to 1,000 times the predicted intensity and then inexplicably dropped to zero. The counts began again at about 10,000 kilometers, only to drop to zero at higher altitudes. This puzzling behavior was repeated many times, especially when the satellite passed over Earth's equatorial regions. One possibility was that the Geiger counters were saturating and ceasing to operate in response to intense radiation. Explorer II, launched on March 26, was equipped to test this hypothesis. Between them the two Explorers yielded enough data that as early as May 1, 1958, James Van Allen announced the discovery of immense doughnut-shaped belts of intense radiation surrounding Earth's magnetic equator. This was the first significant discovery of the space age, and it came as a surprise to everyone.

The solar wind

The solar wind is sometimes considered a discovery of the space age. It is true that the first direct measurements of the composition, velocity, and density of the solar wind were made by space vehicles, including Luniks I and II and Mariner II. However, that interplanetary space is not an ideal vacuum but is occupied by a conducting plasma of ionized hydrogen had been deduced in the years following World War II from a variety of Earth-based observations. Comet tails always sweep from the sun, regardless of whether the comet is approaching or retreating from the sun's vicinity, and it was realized that something stronger than light pressure is required to force back the dust grains and gases of these tails. Moreover, the faint luminosity in the night sky we call the zodiacal light is partially polarized as though it passes through a field of free electrons. These and other lines of evidence led in the late 1950s to the conception of a "solar wind" streaming radially outward from the solar corona.

Space probes now have shown that the solar wind consists chiefly of ionized protons and electrons blowing through the solar system at velocities at 500–800 kilometers per second. In the vicinity of the Earth there are about five protons and five electrons per cubic centimeter. At a distance of about 64,000 kilometers (roughly 10 Earth radii) above the sunward side of the Earth, the solar wind encounters the force of our magnetic field and is deflected around it. The solar wind does not close up again behind the planet. Interplanetary probes have shown that an immensely long magnetic shadow extends from the dark side of the Earth to the far reaches of the solar system. The Pioneer 10 spacecraft, now approaching the vicinity of Uranus, is still encountering it. The shadow may reach to interstellar space, more than 100 times the distance from the sun to the Earth.

Some high-energy protons and electrons do, however, succeed in penetrating the magnetosphere, where they encounter magnetic "crossfire" and are trapped within the Van Allen belts. Within a broad

Figure 10.2
The solar corona, from which the magnetically neutral ionized plasma we call the solar wind streams radially outward through the solar system. (High Altitude Observatory photo)

zone of high radiation, there are two main belts of high intensity. The inner belt, characterized by the presence of high-energy protons, begins at an altitude of about 800 kilometers and reaches peak intensity at about 5,000 kilometers above the Earth's surface. The outer belt, dominated by high-energy electrons, reaches peak intensities at about 16,000 kilometers. A slot of lower intensities lies between the two.

Space probes have provided all our definitive information on the interplay at the unquiet boundary between Earth's magnetic field and the solar wind. This has led to a vastly improved understanding of the ionization and electromagnetic behavior of our atmosphere and of the causes of phenomena such as the polar auroras. Early data from space vehicles also indicated that humans and instruments voyaging through space would have to be shielded from a much higher flux of radiation than had been supposed. Travelers to the moon would pass through the Van Allen belts, and in all likelihood they would face the full glare of the solar wind as they stepped onto the lunar surface. The moon was not expected to have a magnetic field, simply because its density is too low for it to have a large, partially liquid,

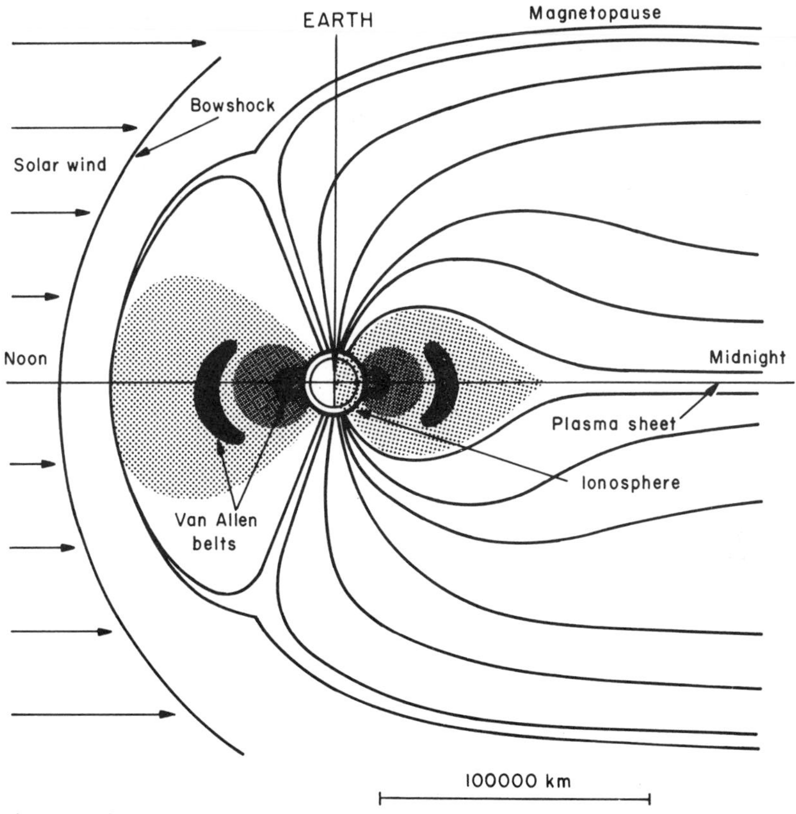

Figure 10.3
Schematic cross section showing the Van Allen radiation belts enclosed within the magnetosphere. From *Orbiting the Sun* by Fred L. Whipple (Harvard University Press, 1981)

metallic core like the Earth's, in which convection currents in the liquid portion generate the field.

The Apollo missions showed that the moon has no magnetic field today; however, some of its ancient crustal rocks show evidence of having formed in the presence of a magnetic field. This surprising observation has not been fully explained. Did the moon originally generate its own magnetic field, which then vanished long ago? Or was the early moon (and its partner, the Earth) briefly subjected to a transient, externally applied field? Whatever the case, the moon is now unshielded from the solar wind and has been for billions of years. Thus, some of the lunar soils are rich in gases implanted by the solar wind, and mineral fragments in certain lunar rocks display radiation damage caused by solar flares more than 4 billion years ago. The radiation effects evident in these lunar materials, and also in meteorites, bear witness to the early history of the sun and the degree to which its radiation has remained constant through geologic time.

The geoid

Vanguard I, the second American satellite, launched on March 17, 1958, confirmed the evidence of Sputniks I and II that the shape of Earth's gravity field needed recalculating. Geodesists represent the shape of the gravity field by a hypothetical figure called the geoid. A gravimeter would register the same value at every point on the geoid, and if seawater could percolate freely through all the rocks of the Earth and attain a level determined solely by the force of gravity the resulting universal sea level would define the geoid.

In the seventeenth century Isaac Newton calculated that the Earth is not quite spherical. The centrifugal force of its rotation produces a slight bulge at the equator and a flattening at the poles. Newton concluded that the Earth is an ellipsoid of revolution in which the polar radii are shorter than the equatorial by one part in 230. By the 1920s we knew from geodetic measurements made on the Earth's surface that the flattening is only about 1/297—considerably less than Newton supposed. The orbits of the earliest satellites indicated that the flattening is closer to 1/298. Then, careful tracking of Vanguard I showed that the flattening is not symmetrical; the distance from the center of the Earth to the south pole is measurably shorter than to the north pole. In short, the gravitational figure of the Earth is pear-shaped. Vanguards II and III added the complication that the equatorial plane of the geoid is not circular but slightly elliptical; the longest

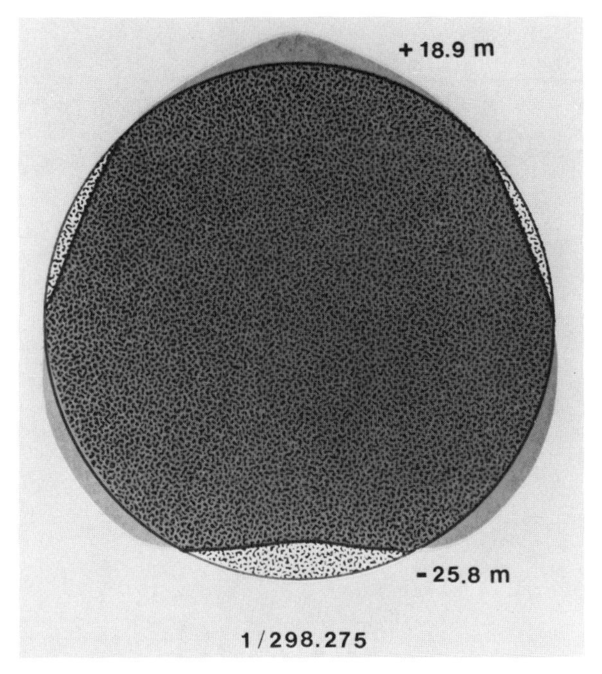

Figure 10.4
The pear-shaped gravitational figure of the Earth, drawn with reference to the mean best-fitting ellipsoid of revolution with a polar to equatorial radius ratio of 1/298.275. (Not to scale)

diameter (between longitudes 32°W and 147°E) is 400 meters longer than the shortest.

After decades of meticulous analysis of the orbits of many satellites, we conclude that the best-fitting ellipsoid of revolution has a polar-to-equatorial axis ratio of 1/298.257. In such a figure, the north pole rises in a hump 18.9 meters high and the south pole sits in a hollow 25.8 meters deep (figure 10.4). There are additional humps and hollows in the body of the geoid, as shown in figure 10.5. Hollows equivalent to −45 meters and −54 meters, respectively, occur near Hudson Bay and in the southwestern Atlantic Ocean, and humps of about +61 and +74 meters over the north central Atlantic Ocean and New Guinea. The greatest and most enigmatic feature in the Earth's gravitational figure is the deep hollow of −104 meters just off the southern tip of India.

All these irregularities in the gravity field reflect an inhomogeneous distribution of mass in the Earth's interior. Interestingly, they do not center in any regular way over continents, ocean basins, or other major

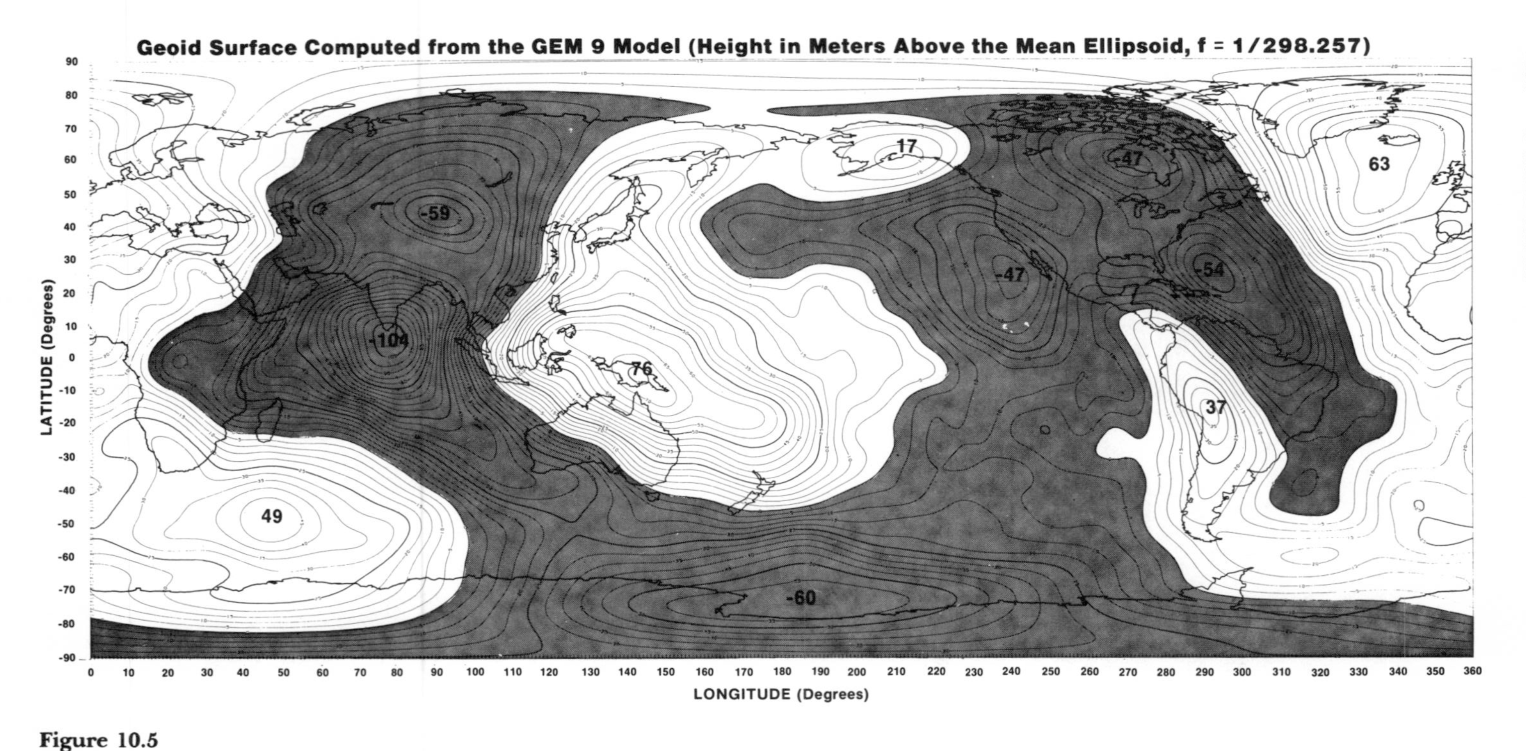

Figure 10.5
The global distribution of humps and hollows in the geoid, as computed from Goddard Earth Model 9 (elevated areas colorless, depressed areas dark). (Courtesy of Goddard Space Flight Center)

features of the observable surface topography. The gravitational effects of the continents and oceans are so fully compensated by a thickening of lighter rocks under the continents and a thinning of denser rocks under the ocean floors that satellites do not "see" them.

What, then, gives rise to these irregularities in the geoid? Several factors may influence the contours of the gravity field. Increasing evidence suggests some of the swells and swales in the geoid correlate with values of heat flowing out through the crust. Some of the most pronounced highs in the geoid occur in regions of relatively high heat flow, and some of the lows in areas of lower heat flow. Enhanced heat flow may mark the site of a rising plume or limb of a convection cell carrying up warm material from deep in the mantle. This material would add to the mass sensed by satellites. Conversely, low values of heat flow may reflect plunging limbs of convection cells that cause a deficiency of mass in the uppermost mantle. Another possible cause that has been suggested for hollows in the geoid is that they mark the wakes of drifting continents. Perhaps the deep hollow near India marks a spot where India lay before its final, rapid drift northward to collide with Asia, and the Earth has not yet compensated for the loss of mass at that site. This suggestion is very tentative, of course, and must not be taken as the answer until much more definitive work has been done on the general causes of geoidal irregularities. We do know, however, that the Earth is not plastic enough to respond immediately to internal changes of mass. For example, the present equatorial bulge is too pronounced for the Earth's present rate of rotation. The bulge is of the right order to correspond with the slightly faster rate at which the Earth was rotating about 15 million years ago. There is, therefore, a considerable lag in the smoothing out of the gravity field after a redistribution of mass takes place or any other form of stress is applied to the planet.

Measurement of crustal motions

Earth scientists are especially concerned with the geoid because, now that the theory of plate tectonics dominates geological thinking, we would like to test it by actual measurements. No one has ever seen a plate move, and until recently it has been impossible to set up any system of measurement and hope to obtain a result during our lifetimes. (I refer here to the relative motions between major land masses or between large regions within land masses; not to motions on opposite sides of individual fault zones, some of which have been measured by ground-based geodetic methods.) Before the space age it was virtually impossible to map accurately sites on the Earth, for all calculations had built-in errors of several hundred meters.

The establishment of worldwide networks of satellite-tracking stations promised that the measurement of crustal motions might become feasible. In 1956 (before Sputnik) it was possible, by the most accurate techniques available, to locate a site on the Earth's surface to a precision of ± 100 meters. By 1966, after 7 years of satellite tracking by special cameras, scientists at the Smithsonian Astrophysical Observatory could locate their stations with a precision of ± 20 meters. At that time they began replacing the cameras with laser ranging systems, and by 1976 the precision had increased beyond ± 1 meter and was approaching ± 10 centimeters.

On May 4, 1976, a new type of satellite specifically designed for laser tracking of high accuracy was launched. Called LAGEOS (LAser GEOdynamic Satellite), it is a small, heavy sphere placed in a very high, stable orbit. LAGEOS carries no electronic equipment of any kind; it is a passive object to be sought out by laser beams. With a diameter of 60 centimeters and a weight of 1,986 kilograms, LAGEOS consists of two precisely machined aluminum hemispheres fitted over a cylinder of solid brass. Its surface is studded with 426 accurately polished and positioned reflectors; 422 of them are of fused silica, which is resistant to degradation by dust and radiation, the other 4 are of germanium. LAGEOS follows a nearly circular orbit, inclined to Earth's equator at an angle of 100°, at an altitude of 5,900 kilometers. At such a height, LAGEOS is visible from all points on Earth and is beyond the effects of atmospheric or gravitational drag. Laser reflection measurements to this satellite, and to certain others, can achieve a precision of 1–3 centimeters once problems caused by atmospheric effects are overcome.

LAGEOS is expected to remain in orbit for about 10 million years. With such a life expectancy, the satellite seemed an appropriate vehicle to carry a message to our descendants, who might retrieve it from space or observe its reentry millions of years hence. Accordingly, a message designed by astronomer Carl Sagan was inscribed on a thin sheet of stainless steel and wrapped around the brass cylinder. The message gives the name of the satellite in English, just in case people still speak or read English, and in binary. Earth is shown orbiting the sun, and LAGEOS orbiting Earth. Three maps show the continents as they are today, as they were about 225 million years ago, and as we think they will be (if plate motion proceeds at an average rate of 2.5 centimeters per year) about 8 million years in the future.

Whatever role LAGEOS may play in the distant future, today it is serving its purpose as an extremely reliable laser reflector. Networks of tracking stations have begun collecting long-term records of measurements from LAGEOS and other satellites in order to detect plate motion. The laser ranging results are sometimes combined with those of a completely different measurement technique called Very Long Baseline Interferometry (VLBI), which was described in the preceding chapter. VLBI stations are usually dedicated to the receipt of radio signals from quasars or other astronomical objects. However, by the very accurate timing of signal reception, stations that are widely separated (on different continents or opposite sides of continents) can locate themselves with respect to each other with a precision of 3–6 centimeters. When they overcome the problem of pathlength changes due to water vapor in the atmosphere, VLBI scientists expect to achieve a precision of 1–3 centimeters. (Distances are measured not over the curvature of the Earth but along a chord between VLBI stations.) Clearly, the technology now exists for measuring the relative motion of the Earth's crustal plates, if these motions are proceeding at the estimated rates of 1–10 centimeters per year.

Repeated observations, made for several years from certain stations, already have begun to yield the first tentative indications of such crustal motion. A Smithsonian laser station in the Andes at Arequipa, Peru, reports that the ground under it appears to have been rising at the rate of 4 centimeters per year for the past six years. Other observing stations at San Diego and at Quincy, California, appear to be approaching each other at a rate of about 9 centimeters per year. The area between these two sites is crisscrossed by fault zones; however, no structural analysis predicted a crustal shortening at so high a rate. Geologists conclude that if these measurements are correct, we may have some drastic surprises ahead as we try to sort out the directions and rates of plate motion in that area. In contrast with these readings, VLBI measurements show essentially zero deformation (less than a centimeter per year) between stations in Massachusetts and the Owens Valley of California. An equivalent lack of intracontinental motion has been recorded across Australia. With many satellite-ranging and VLBI stations now being positioned around the world, we can expect the 1980s to bring us the first definitive data on the instability of the Earth's crust.

Not only do certain artificial satellites carry laser reflectors, so does the moon. Reflectors were placed on the lunar surface during four Apollo missions and two of the Soviet Union's unmanned Luna missions. The Lunar Laser Ranging System will soon be capable of detecting crustal motions. Already, more than a decade of measurements have yielded precise new data on the orbital and rotational dynamics of the moon and the Earth and new information bearing on theories of general relativity.

Weather satellites

On April 1, 1960, the United States launched TIROS I, the first experimental weather satellite. Ten days later the city of Brisbane, Australia, was alerted to the presence of a typhoon 800 miles offshore in a remote area where observations were not ordinarily available. Since then, nine more TIROS satellites and several other series, including NIMBUS and NOAA, have gone into orbit. The data they provide have revolutionized not only our ability to predict short-term weather systems, but also our understanding of the effects of the sun on our atmosphere.

The weather satellites are equipped with both optical-band video cameras and infrared detectors, and thus are capable not only of imaging cloud cover and snow and ice distributions but also of measuring the temperature of the atmosphere and oceans. They constantly monitor the heat balance between the absorption of sunlight and the emission of infrared radiation over the entire surface of the Earth.

On the day it was launched, the first NIMBUS satellite transmitted pictures of an important but previously unknown type of cloud pattern, consisting of a shingle-like distribution of large cells each of which is 30–80 kilometers across. The cells are so wide that no more than one of them at a time was ever observed from aircraft. Figure 10.6 shows a canopy of these clouds over the northern Pacific Ocean.

From the first, weather-satellite data have been correlated with those gathered by aircraft that fly into storms to record temperatures, pressures, humidity, and wind dynamics. Experiments with cloud seeding to generate rain or reduce the force of hurricanes have been carefully observed and their effects traced over wide areas. Since 1960 the weather satellites have identified and tracked every major storm. Today they relay information every half-hour to the National Hurricane Center in Miami and the Joint Typhoon Warning Center in Guam. The saving of lives over the past two decades has been tremendous.

In addition to short-range weather forecasting, the satellites have located previously unrecognized areas of storm formation. One of these lies in the Atlantic Ocean off the west coast of Africa, another off the west coast of Mexico. Observers have learned to spot rising columns of warm air in such regions and predict their evolution into

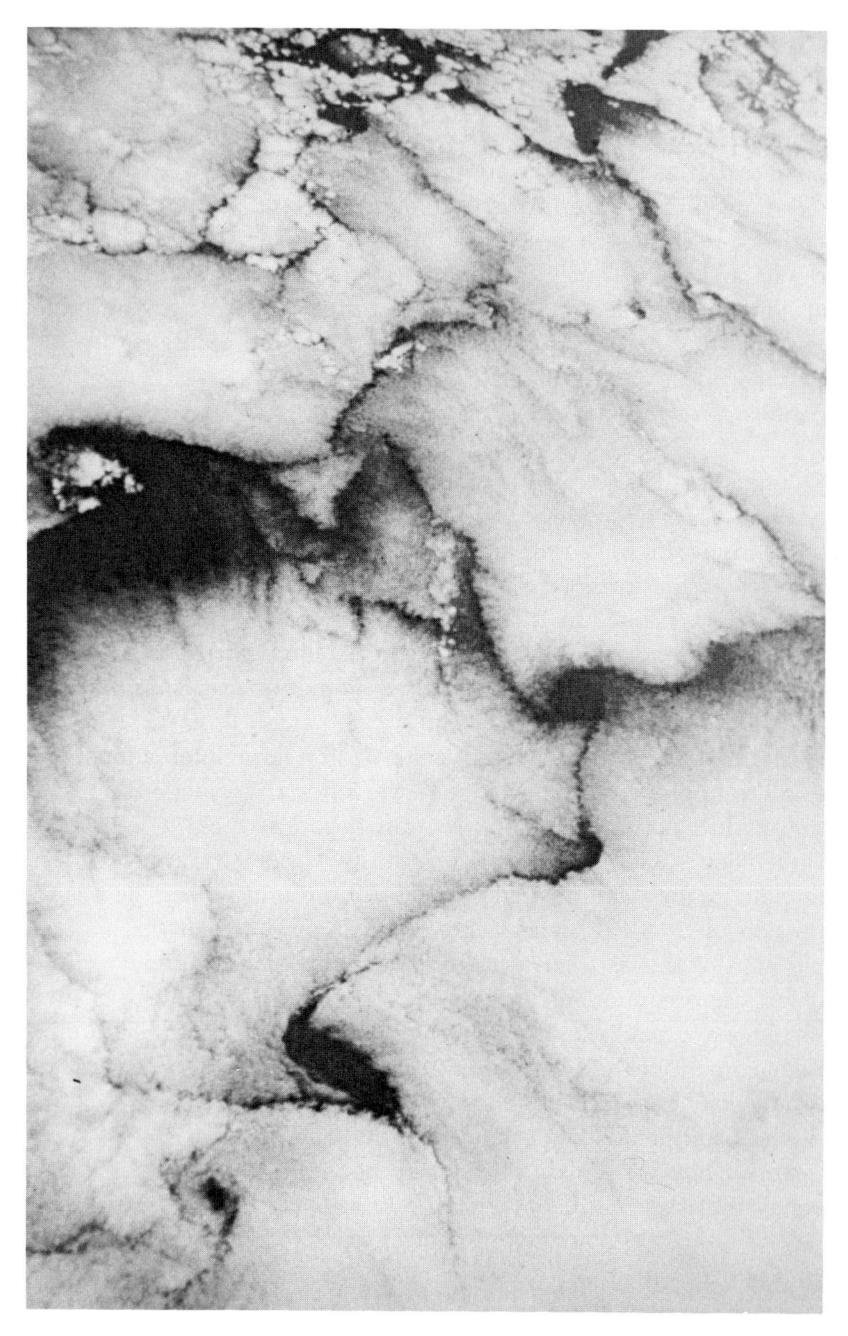

Figure 10.6
Cloud pattern discovered by Nimbus I. Individual cells in this shingle-like array may be as wide as 80 kilometers. (Courtesy of NASA)

storm centers even before characteristic clouds have formed. At high latitudes, the satellites monitor the distribution of ice floes in shipping lanes. In dry areas they track dust storms and even swarms of locusts as they follow air currents over the ground. Many agencies use weather-satellite data as a guide in water management or flood control. Among the most impressive symbols of modern life are the presentations on network television of time-lapse images showing the distribution of clouds and the advance of frontal systems over the entire continent of North America. Weather satellites have introduced an awareness of the atmosphere and its global patterns into our everyday lives.

Earth-resources satellites

The first Earth Resources Technology Satellite (ERTS 1) was launched in 1972. A second followed in 1975; at that time the name was changed to Landsat. (ERTS 1 retroactively became Landsat I, but a host of pictures and reports were published bearing the ERTS title.) A third Landsat was orbited in 1978, and one or two more are scheduled for the 1980s. In addition, the Skylab space station carried an Earth Resources Experiment Package that included optical, infrared, and microwave sensors.

The Earth-resources satellites are part of a large international program to improve our knowledge of the lands and waters that sustain the world's population. The program began in the 1960s with the United States and about 50 other countries working toward correlating satellite images with air photographs and ground surveys. It was the mission of NASA to develop the technology; dissemination and application of the data were to be the responsibility of other agencies or private organizations. Images and tapes from the Earth-resources satellites are available to investigators around the world. The future management of the program is still under discussion; meanwhile the productivity has been immense.

The Landsats follow near-polar orbits at altitudes of 900–920 kilometers. They are timed to be synchronous with the sun, crossing the equator 14 times a day at about 9:30 A.M. local time. Nearly every part of the globe is imaged by a given satellite every 18 days, so it is possible, as with the weather satellites, to monitor changes in crop distribution, snow cover, river flooding, or air and water pollution.

Landsat images are not photographs. They are produced by multispectral scanners which store information in digital form that is resolved into images. These images are published in different colors depending on which single or multiple channels have been selected to serve some special line of research.

The Landsats orbit at such high altitudes that their images are undistorted. Each one overlaps its neighbors by about 15° at the equator and 85° in the polar regions. The resolution is normally 80 meters, although certain linear features (such as roads or canals) that are only 10 meters wide may stand out clearly against contrasting backgrounds. By matching adjacent images it has been possible to create mosaics of very large areas, such as the first-of-a-kind "picture" of the lower 48 states.

Many a map has needed alteration since Landsat images revealed the actual distribution of lakes, canyons, mountain masses, or shoreline features. Geologists have also discovered structural features that are too large but too subtle to be recognized on the ground or encompassed within aerial photographs. Thus, completely new information has been developed on faults and fracture systems that are good prospecting sites for mineral deposits or sources of fresh water and on folds and domes that may be oil or gas traps. Environmentalists use Landsat images to search out safe routes for pipelines, to monitor the progress of strip mining, and to measure the effects of waste disposal and other natural and man-made processes; agriculturists use them to predict crop yields. There are many other applications. (Books have been compiled of beautiful and informative Landsat images. Two sources are listed at the end of this chapter for readers who may wish to devote hours to happy browsing through Earth Resources Technology Satellite pictures.)

Many other types of satellites have been sent into orbit to measure some facet of the Earth's character. Magsat has mapped the global distribution of magnetic anomalies; Seasat has measured the mean contours of the ocean surfaces and found that they reflect major features of ocean-floor topography. Yet, for all their sophistication, these remote sensors reveal only the Earth's upper layer. They cannot tell us how and when the Earth formed, nor what the starting materials were, nor what the Earth's bulk composition is. These are among the most critical things we would like to know about our planet, but we could only guess at most of the answers if we were to focus all our attention on the Earth itself. Ironically, the space age has taught us that to learn more about this world we must study extraterrestrial examples and samples.

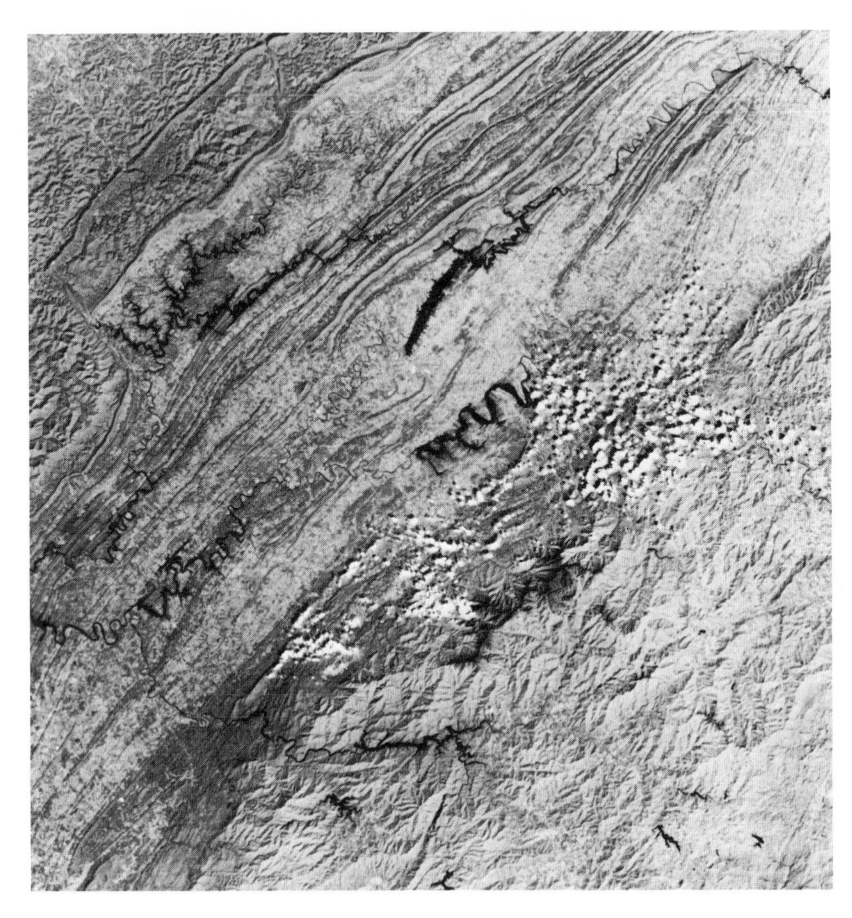

Figure 10.7
The plunging folds of the Appalachian valley and ridge province in the vicinity of Knoxville, Tennessee. The Great Smoky Mountains, which dominate the region, look rather inconspicuous at center right. (Landsat I image)

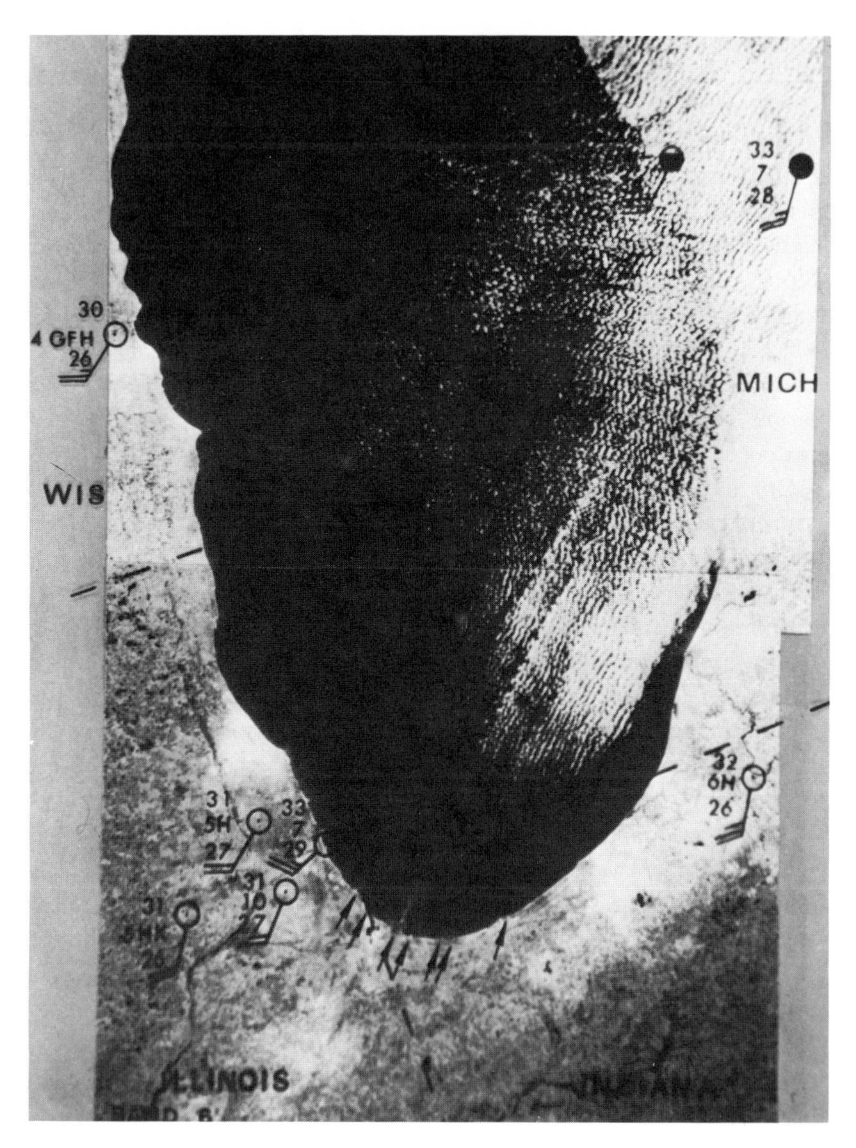

Figure 10.8
Mosaic of the southern part of Lake Michigan marked with wind direction and velocity symbols. (Landsat images)

The composition and the age of the Earth: Clues from meteorites

The Earth has a radius of 6,371 kilometers, but we can examine only its outermost skin. The deepest caves in the world reach down only about 1 kilometer; the deepest drill holes, bored in the search for oil, penetrate about 6 kilometers. Here and there on the continents are diatremes, narrow pipes of material that rose explosively from great depths, bringing up fragments of the upper mantle. Some of these fragments, or mantle xenoliths, probably originated at depths of about 100 kilometers. If so, that means that the deepest materials we ever examine come from the uppermost 1–2 percent of the radius. The nature of the deeper 98 percent must be deduced from indirect evidence.

Our information on the Earth's interior now comes from several types of observations. As seismic waves travel through the Earth they undergo abrupt changes of velocity and pathway that mark boundaries between layers of different densities. The three main domains are the crust, mantle, and the core. Seismic waves give evidence of the densities in the interior but tell us nothing specific about the composition of the materials they pass through. To build a credible picture of the chemical and mineral composition of the interior, we analyze the waters, vapors, and molten rocks of volcanic eruptions and perform experiments on natural and artificial rocks to observe their behavior under conditions simulating the temperatures and pressures at various depths. In recent years, instruments called diamond anvils have been developed that can subject materials to static pressures up to a million bars, equal to those at the core-mantle boundary. These experiments produce reasonable facsimiles of the materials that may actually be present within the Earth. Still, because we cannot obtain samples of the deep interior, some scientists examine the lunar soil for links to Earth, others seek the evolutionary record in the histories of our companion planets, and still others look for analogous materials among the meteorites that fall from space.

Meteorites are of three main types: iron, stony, and stony iron. About 4 percent of all meteorites are irons, which consist almost entirely of metallic nickel-iron. Stony irons, which consist of half metal and half stony matter, make up about 2 percent of meteorites. The remaining 94 percent are stony meteorites. By all odds, the most abundant type of stony meteorites are the so-called ordinary chondrites, which are fragmental aggregates of silicate minerals, metal grains, and chondrules (small spherical bodies of silicate minerals and glass having no counterparts among terrestrial rocks). The ordinary chondrites have

Figure 10.9
An iron meteorite discovered on a high plateau of windswept ice near the Allan Hills of Victoria Land, Antarctica, in January 1982. The hand-held device shows a 6-centimeter scale and the field number of the meteorite. (Photo by author)

Figure 10.10
An ordinary chondrite, partly covered with fusion crust, found in the Allan Hills region of Antarctica in 1979. (NASA photo)

never been melted, but some of them have been heated and recrystallized at temperatures below the melting point.

Meteorites yield important information on the age and composition of the solar system, and hence of the Earth. Isotopic age determinations (measurements of radioactivities in certain elements) show that the oldest rocks yet discovered in the Earth's crust formed about 3.8 billion years ago. The planet itself, however, formed earlier than that—how much earlier was not determined until the 1950s, when ages of terrestrial rocks and of meteorites were found to extrapolate back to a common time of origin about 4.6 billion years ago. Therefore, 4.6 billion years is taken as the age of the solar system. Recent measurements on lunar rocks show that they, too, extrapolate back to the same time.

Two relatively rare but important varieties of stony meteorites are achondrites and carbonaceous chondrites. The achondrites are igneous rocks that appear to have originated in extraterrestrial lava flows. Carbonaceous chondrites are fragmental aggregates with bulk compositions closely resembling the abundance of elements (other than gases) that have been measured in spectra of the sun. These meteorites have all been altered somewhat during their long histories. Nevertheless, they are the most ancient and primitive materials available to us. To

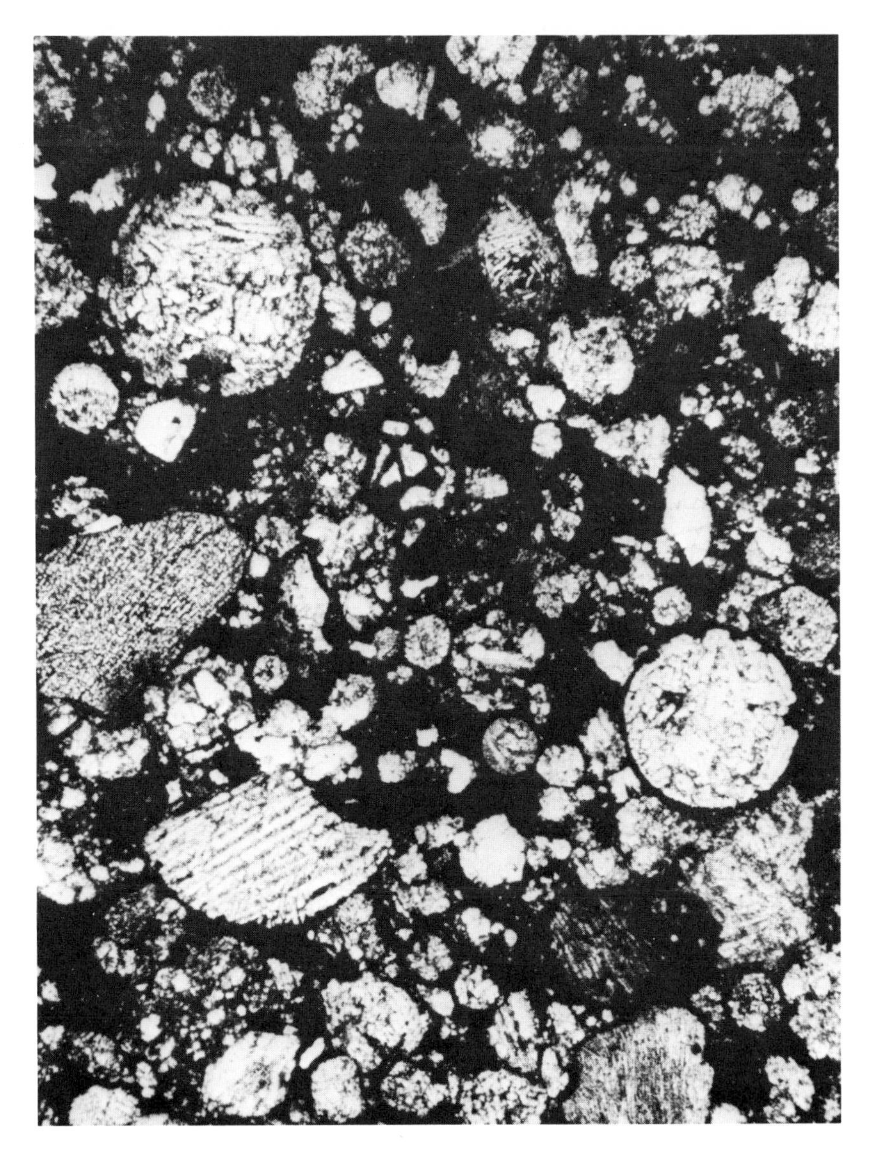

Figure 10.11
The fragmental texture of a chondritic meteorite, as seen under a microscope. Spherical chondrules are admixed with mineral grains and angular chondrule fragments embedded in a fine-grained matrix. (Magnification 40×; the two largest chondrules are about 1 millimeter in diameter.) (Smithsonian Astrophysical Observatory photo)

pick one up is almost like holding a sample of the aggregated dust in the original nebula from which the sun and all its planets and satellites formed.

The most interesting components of the carbonaceous chondrites are organic molecules, including amino acids and fatty acids, which may have been present in the accreting nebula or may have formed after incorporation into the meteorite parent bodies. Their optical properties indicate that these compounds are of nonbiological origin. To date, no extraterrestrial samples have shown positive evidence of life. Still, the meteoritic compounds are of special importance because they show what kinds of hydrocarbon molecules existed before the advent of life, which, according to present age determinations, first appeared on Earth about 3.5 billion years ago.

The bulk composition of the Earth may well approximate that of chondritic meteorites, but, as noted above, the planet is now separated into layers of differing densities and compositions. Looking at the varieties of meteorites, scientists began in the middle of the nineteenth century to speculate that meteorites were fragments of a layered planet (or planets) similar in many respects to Earth. One of the earliest models of a "meteorite planet" consisted of a core of nickel-iron metal, a transition zone of stony iron, a mantle of chondritic stone, and surficial layers of achondritic lavas.

Today we are convinced that there was never a single meteorite parent planet. We believe instead that meteorites represent the collisional debris of many small asteroidal-size bodies, most with diameters of less than about 1,000 kilometers. Some meteoritic minerals are of a type that can occur only in low-pressure environments; none show evidence of having formed under high pressures such as those in the deep interiors of bodies as large as our moon.

We no longer imagine that we would find materials with the mineralogy and textures of stony meteorites if we could sample the Earth's mantle. We do, however, still think of the Earth's core as roughly analogous to iron meteorites. The Earth clearly has a dense core that gives rise to a magnetic field, and iron is among the most abundant of the heavy elements. The comparison cannot be exact, however, because the core is slightly less dense than nickel-iron metal and so must contain a small proportion of a lighter element. Controversy has raged over whether that element is likely to be silicon, potassium, or sulfur.

Within the past three decades, photographs have been taken of the incandescent trails of several incoming meteorites. The orbits calculated from the films show that the bodies followed elliptical paths that carried them around the sun from the asteroid belt between Mars and Jupiter. We conclude, therefore, that collisions between asteriods must send fragments of silicate rock and metal into orbits that cross the path of the Earth. The resulting impacts on our planet can be events of major geological significance.

The impact of meteorites on the moon and the Earth

Meteorites fall to the Earth every day, but most of them decelerate in the atmosphere and plummet to the surface at the velocity of free fall. These bodies may plunge through roofs (rarely) or make pits in the soil, but they seldom cause much damage. Indeed, so little of the Earth's surface is populated that it is a rare and wonderful event even to witness a meteorite fall.

However, certain scars in the landscape have been recognized as the results of explosive meteorite impacts. This process clearly involved energies that are orders of magnitude greater than those of free fall. Calculations show that bodies weighing 10 tons or more can pass unhindered through the atmosphere and strike the Earth at cosmic velocities. The collisions instantaneously vaporize the projectile, melt and metamorphose the target rock, blast open a crater, and blanket the countryside with fragmental ejecta. Shock waves traveling through the ground uplift and tilt the country rock and, if the event is energetic enough, surround the site with a series of concentric rings of solid rock. Comparisons between the structures and rock textures at suspected impact sites and those at nuclear-bomb test sites reveal an array of shock-produced features common to both. Using these as criteria, scientists have identified nearly 200 impact scars. Such scars are visible on every continent except Antarctica, and they probably are present there beneath the thick ice.

Much research on impact sites was done in the 1960s. Nevertheless, most geologists continued to regard meteorite impact structures as natural curiosities of no fundamental significance to Earth science until the Apollo missions. Among the rock samples brought back by the astronauts were some shocked rocks that could be dated isotopically. When the results were pieced together with other types of evidence, a picture emerged in which virtually all the lunar craters, from millimeter-size pits to vast multiringed basins more than 1,000 kilometers in diameter, were formed by meteorite impact. Age determinations showed that the great basins were excavated during a bombardment by projectiles between 4.2 and 3.9 billion years ago.

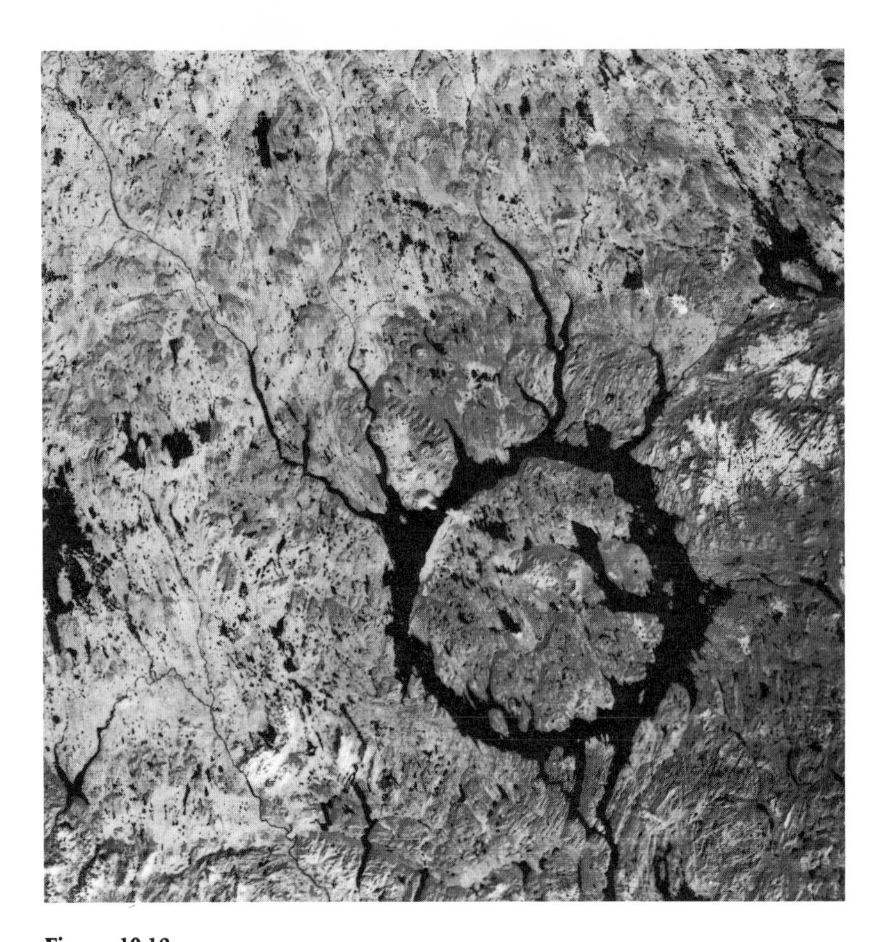

Figure 10.12
This ring structure at Manicouagan, Quebec, is 65 kilometers in diameter. The bedrock bears the imprint of shock metamorphism caused by a hypervelocity impact that occurred 214 million years ago during the Triassic period. Skylab and Landsat images showed that the ring (which is flooded by water impounded by a hydroelectric dam) is surrounded by a circumferential depression about 145 kilometers in diameter. (Landsat image)

Later, between 3.8 and 3.1 billion years ago, these basins were flooded with dark basaltic lava.

Isotopic dating also indicated that the Earth and the moon are equally old and appear to have been linked together from the beginning. If the moon underwent a great bombardment about 4 billion years ago, so too did the Earth. But the Earth apparently has no relics of crust older than 3.8 billion years. All traces of that catastrophic period are lost. Or are they? Suddenly, we realize that the cratered face of the moon provides us with a well-preserved record of what happened during part of the 800-million-year missing interval of Earth's history—the period between the formation of the planet 4.6 billion years ago and the emplacement of our oldest continental rocks.

Clearly the Earth and the moon have had very different evolutionary histories. The composition of the light-colored lunar crustal rocks indicates that the nascent moon was covered by a deep ocean of molten magma. But the crustal rocks cooled early, and the entire moon was rigid enough within the first 400 million years to hold open huge basins in the crust when the bombardment began. As similar projectiles plunged into our own much larger and more thoroughly melted planet, they must have destroyed all traces of any thin crust that had begun to form. By penetrating deep into the body of the Earth they may have contributed significantly to the inhomogeneity of the mantle. They may even have influenced the locations at which future continental rocks would begin to grow. None of this is certain, but our new knowledge of the moon has generated these novel lines of speculation.

Since 1980, interest in meteorite impact has intensified as a result of published analyses showing that iridium is enriched in a thin layer of clay, 65 million years old, that separates the strata of the Cretaceous and Tertiary periods. Iridium is not the only component; several other elements that are rare in terrestrial rocks but characteristic of meteorites also occur in the clay, and in cosmic proportions. These elements constitute strong evidence that an impact of very large scale raised immense clouds of meteorite-contaminated dust that settled around the world.

But the story does not end there. In an imaginative intellectual leap, the scientists who discovered the iridium enrichment proposed that the impact caused the wholesale extinctions of dinosaurs and many other forms of life (including every vertebrate weighing more than 25 kilograms) at the end of the Cretaceous period. Some proponents of this theory argue that the projectile struck on land and raised so much fine dust into the stratosphere that darkness ensued, photosynthesis declined, and the food chain that sustained the large vertebrates was

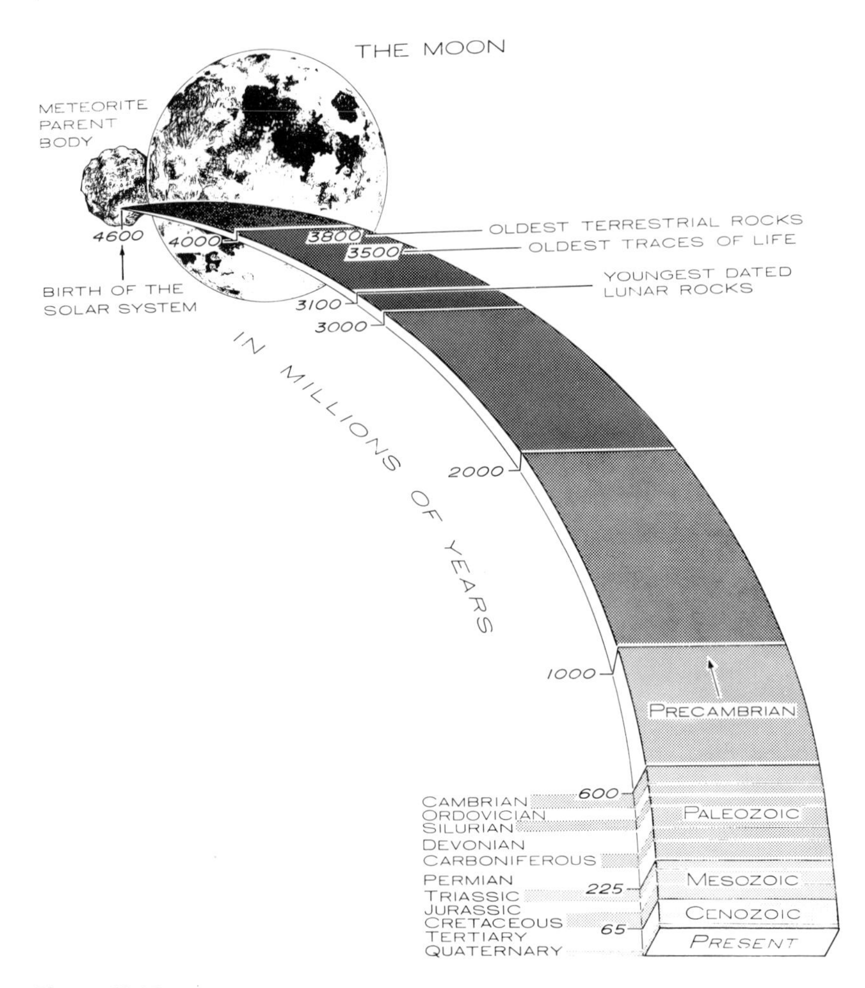

Figure 10.13
The geological time scale. The moon and a meteorite parent body, two sources of rocks more ancient than any yet discovered in the Earth's crust, are shown in their proper time frame. (Harvard-Smithsonian Center for Astrophysics illustration by Beryl Langer from the author's design)

Figure 10.14
(By permission of Johnny Hart and Field Enterprises, Inc.)

Figure 10.15
Earth in space, photographed by the Apollo 17 astronauts. (Courtesy of NASA)

destroyed; others suggest that the projectile plunged into the ocean and sent so much water vapor into the stratosphere that a "greenhouse effect" overheated the atmosphere and stifled living things. Recent calculations show that an oceanic impact could also have drowned huge populations of land animals by raising a great wave that washed over low-lying lands; however, paleobotanists have examined the Cretaceous-Tertiary boundary and found no evidence of a break in the succession of leafy plants, and some paleontologists argue that the dinosaurs were not snuffed out suddenly but succumbed gradually over a long period. Geochemists are now examining other major breaks in the fossil record for enrichments in meteoritic elements. Obviously, the issue is far from settled, and persuasive evidence, pro or con, may be a long time in coming. Meanwhile, the excitement raised by this controversy is infectious. At last, Earth scientists of many disciplines are looking to the solar system to better understand the history of the Earth.

Seeing the whole Earth

The most valuable of all the contributions of the space age may be the views of the whole Earth suspended in the blackness of space. One astronaut described the Earth as looking fragile. That vision seems to have affected many of us when we saw, for the first time in history, the isolation of our world and the apparent vulnerability of its thin film of living matter. The first full Earth pictures were published in 1967. The concept of "Spaceship Earth" and the impetus for conservation and protection of our environment burst into full flower in 1968. If the space program was the cause of this, and I believe it was, then it has repaid its costs many times over.

Reading

Short, Nicholas M., Paul D. Lowman, Jr., and Stanley C. Freden. Mission to Earth: LANDSAT Views the World. NASA report SP 360. Washington, D.C.: Government Printing Office, 1976.

Sullivan Walter. *Continents in Motion: The New Earth Debate.* New York: McGraw-Hill, 1974.

Whipple, Fred L. *Orbiting the Sun: Planets and Satellites of the Solar System.* Cambridge, Mass.: Harvard University Press, 1981.

Williams, Richard S., Jr., and William D. Carter, eds. ERTS-1: A New Window on Our Planet. U.S. Geological Survey professional paper 929. Washington, D.C.: Government Printing Office, 1976.

Contributors

Leo Goldberg is Higgins Professor of Astronomy, Emeritus, at Harvard University.

James W. Head III is Professor of Geological Sciences at Brown University and a member of the imaging team for the NASA Galileo mission to Jupiter, and directs the NASA–Brown University Regional Planetary Data Center.

John A. Wood is Professor of the Practice of Geology at Harvard University, a geologist at the Smithsonian Astrophysical Observatory, Associate Director for Planetary Sciences at the Harvard-Smithsonian Center for Astrophysics, and head of the Center's Extraterrestrial Petrology and Mineralogy Group.

Bradford A. Smith is Associate Professor of Planetary Sciences at the University of Arizona and leader of the Voyager Imaging Team.

Randolph H. Levine is Manager of Research and Development for Computer-Based Education at the Digital Equipment Corporation.

Andrea K. Dupree is an astrophysicist at the Smithsonian Astrophysical Observatory and the Associate Director for Solar and Stellar Physics at the Harvard-Smithsonian Center for Astrophysics.

Jonathan E. Grindlay is Professor of Astronomy at Harvard University and an astrophysicist at the Smithsonian Astrophysical Observatory.

Paul Gorenstein is an astrophysicist at the Smithsonian Astrophysical Observatory, a Lecturer on astronomy at Harvard University, and the principal investigator for the Large Area Modular Array of x-ray telescopes being developed at the Center for Astrophysics.

George B. Field is Professor of Astronomy at Harvard University and a Senior Scientist at the Smithsonian Astrophysical Observatory.

Ursula B. Marvin is a geologist at the Smithsonian Astrophysical Observatory and a Lecturer in the Department of Geological Sciences at Harvard University.

James Cornell is Publications Manager at the Harvard-Smithsonian Center for Astrophysics.

Index

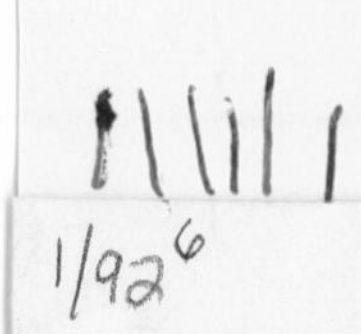